S0-CKS-229

Regional Studies

The Palisades Sill, New Jersey: A Reinvestigation

Kenneth R. Walker

THE
GEOLOGICAL SOCIETY
OF AMERICA

SPECIAL PAPER 111

The Palisades Sill, New Jersey: A Reinvestigation

Kenneth R. Walker

Bureau of Mineral Resources, Canberra City, Australia

THE
GEOLOGICAL SOCIETY
OF AMERICA

SPECIAL PAPER 111

Library of Congress Catalog Card Number: 70-90031
S.B.N. 8137-2111-3

Published by
THE GEOLOGICAL SOCIETY OF AMERICA, INC.
Colorado Building, P. O. Box 1719
Boulder, Colorado 80302

Printed in the United States of America

The printing of this volume has been made possible through the bequest of Richard Alexander Fullerton Penrose, Jr., and is partially supported by a grant from The National Science Foundation.

PREFACE

A summary of this Special Paper on the Palisades Sill will appear in Memoir 115 of The Geological Society of America, a volume of papers in tribute to the late Professor Arie Poldervaart. The length of the present paper prohibited its full appearance in the memorial volume, but it is intended here as an extension of that tribute. Kenneth Walker expressed this feeling eloquently in a recent letter ". . . it was my wish to dedicate the work to Arie in tribute and admiration of his unstinting interest and devotion to his science and his students, and as an expression of my appreciation for the stimulating encouragement he gave me while undertaking an investigation so close to his fundamental interests, an experience that endeared my respect for him as a great scientist."

Poldervaart's death was tragically early and painful. He was thus denied the full fruits of his own hard labor which had led him toward grand and challenging ideas on the early history of the earth. He was similarly denied a view of the growth of his students and colleagues and of the flowering of ideas he shared with them. To have seen this paper of Walker's on the Palisades Sill would have given him immense satisfaction, since he had long believed it should be reinvestigated, with great attention to details of mineralogy, chemistry, and petrology. Probably no other paper, therefore, could more fittingly honor the memory of Arie Poldervaart.

Leonard E. Larsen
(Editor of Poldervaart Volume, 1969)

ACKNOWLEDGMENTS

The laboratory work for the investigation was carried out at Columbia University, Yale University, and the Australian National University. I am deeply indebted to many in these Universities for stimulating discussion, but in particular I wish to record my sincere appreciation to the late Professor A. Poldervaart, discussion with whom stimulated and inspired me during the early stages of the investigation.

I also wish to thank Professors K. K. Turekian and S. R. Taylor (both of whom familiarized me with the optical emission analytical procedures used for the most part in this work), and Dr. W. Broecker, for the use of facilities in their geochemical laboratories. The Bureau of Mineral Resources kindly gave me the opportunity to use their laboratory facilities at convenient times for petrographic determinations and to check analytical work. The Bureau also gave much assistance with drafting and typing. The approval of the Director of the Bureau of Mineral Resources to publish is acknowledged.

Dr. K. Lowe made available a map and samples of diamond drill core from holes 14, 18, and 19 at Haverstraw Quarry, which have proved invaluable in this work. Professor H. H. Hess provided me with unpublished analyses of chilled dolerite of the Palisades Sill, and also with two analyzed plagioclases, one from the Stillwater and the other from the Bushveld intrusion, for checking the accuracy of the method used for plagioclase determinations. T. Nicholas checked R.I. determinations made on the plagioclase glasses. I. Pontifex examined three polished sections and reported on their opaque mineral content. K. Zawartko assisted with the photomicrographs. Dr. D. H. Green made the electron probe X-ray microanalyses of olivines reported in this paper. The help from these people is greatly appreciated.

I am grateful for the company of the late Professor Poldervaart, Professors F. Donath and J. C. Jaeger, and Dr. K. Lowe on various days while doing field work. Dr. A. Parker kindly assisted with specimen collections of the Edgewater and Union City Sections. I also gratefully acknowledge the benefit derived from the critical reading and discussion of the manuscript by Professors A. E. Ringwood, S. R. Taylor; Drs. D. H. Green, K. S. Heier, J. A. McDonald, I. McDougall, J. F. Lovering, S. A. Morse, and A. J. R. White; S. E. Smith and K. A. Townley.

I wish to thank the Australian National University and Columbia University for financial support through scholarship and research funds; and also the United States Education Foundation for the help afforded through a Fulbright grant.

Contents

FIGURES

PLATES

TABLES

ABSTRACT

The Palisades Sill of New Jersey and New York states was emplaced in the Newark Formation in the Upper Triassic, and is about 1000 feet thick. Its form is typical of a hypabyssal sheet except for some dike-like features, which are mainly in the northern part of the intrusion. The sill is a multiple intrusion in which two magma phases of oversaturated tholeiite have been recognized. Olivine crystallized in appreciable amounts only after the emplacement of the second and larger phase. The accumulation of olivine enveloped in early plagioclase and augite determined the location of the famous hyalosiderite dolerite layer above the crystallized part of the first phase. The order in which the essential minerals began to crystallize in each magma phase was olivine, plagioclase, augite, and orthopyroxene. In the northern part of the intrusion, where the Mg-olivine layer is absent, the contact between the two magmas is marked by a reversal in the pyroxene fractionation trend. Equilibrium was established between the two magma phases early in the fractionation sequence. Fractional crystallization dominated the differentiation process, and progressive crystallization proceeded on a normal course to an advanced stage of iron-, silica-, and alkali-enrichment where fayalite granophyre and granophyric dolerite formed.

The whole-rock analyses for the major and minor elements, and for B, Nd, Nb, Pb, and Sn, are presented. In addition, element partitioning and the behavior of elements with fractionation have been determined in the sill by studying the variations of Ca, Sr, Ba, Fe, Mn, Ti, Cr, V, Sc, Mg, Co, Ni, Y, Zr, Mo, La, Cu, and Ga in the rock and mineral series. The elements distribute and arrange themselves in crystal lattice sites during fractional crystallization according to their total chemical properties. It is not only the properties of the ion in the liquid, but also its properties within the crystal lattice, which determine its behavior, and the trace cations like the major cations show varying site preference according to their bonding characteristics in different lattice structures.

INTRODUCTION

The valuable work of Lewis (1907, 1908a, 1908b) on the Palisades Sill established conclusively that the exposure is a complete section of a differentiated basic intrusion. Since then, the Palisades Sill has attracted the attention of many geologists seeking an understanding of the process of differentiation. Many eminent geologists (Sosman, Bowen, Fenner, Daly, Bailey, Barth, Walker, and others) have added their authority to establish the intrusion as a type example of gravitational differentiation, and as such the sill is widely quoted in many petrological textbooks. The history of petrological thought on the intrusion is clearly summarized by Walker (1940, p. 1079-1083). Though it was Lewis who made the first important fundamental petrological observations, it was Walker (1940) in a detailed petrological study, who reaffirmed that the settling of solid phases was the most important factor in the differentiation of the intrusion. However, petrologists have, from time to time, felt that a number of aspects of the petrology of the Palisades Sill remain unsatisfactorily explained, and it appears that Hess (1956) was the first to express in print his doubts about the validity of long-accepted views. But no further detailed observations have been made on the intrusion since Walker's study in 1940. With the advance of petrological knowledge since then, it seemed opportune to re-examine the sill in an attempt to resolve the yet unanswered problems arising from previous studies. During 1961, I had the opportunity to examine the Palisades Sill and collect suites of specimens. Initial inspection revealed new evidence important in understanding the geology of the intrusion, and so I decided to undertake a complete re-examination, paying particular attention to chemistry and to horizons in those parts of the intrusion which have received little attention in the past. I have referred briefly to this investigation previously, Poldervaart and Walker (1962) and Walker (1965), where I indicated that the new evidence would lead to a revision of concepts on the petrogenesis of the sill.

Both Lewis (1907, 1908a) and Walker (1940) provide important background information to the present study, and the reader is referred in particular to the micrometric, grainsize, and specific gravity data given by Walker and to his detailed comments on schlieren and late-stage hydrothermal products (*see also* Walker, 1953). Additional observations of this type in the present study are limited to those required for new chemical and petrological work, and to where the need arose to review previous determinations. Again to avoid the repetition of previous work and the unnecessary enlargement of the paper that follows, comparisons with other intrusions have been limited mainly to where they contribute to argument or to where they are necessary to review the current status of a problem. I have endeavored to present new information in a form so that the reader may, if he wishes, conveniently make further comparisons.

The present paper reports the work of field and detailed petrological studies, including the chemistry of the intrusion. However, the following problems are still receiving attention because of their importance to an understanding of the petrogenesis; they will be reported on in more detail when complete.

(1) The conditions of formation of the Mg-olivine layer and the occluded internal chilled contact within the Palisades Sill, New Jersey (K. R. Walker and D. H. Green, in prep.).

(2) Variations with fractionation in the monoclinic pyroxenes, and, in particular, the relationships between ferroaugite, ferro-orthopyroxene, and fayalite in the late fractionation stages (K. R. Walker and J. F. Lovering, in prep.).

(3) A study of the opaque iron minerals and the effect of opaque inclusions on the chemistry of silicate mineral phases, particularly on the Cr, Ni, and V content of the pyroxenes.

The sill has been referred to previously as the Palisade diabase by Walker (1940). To avoid ambiguity in petrological terminology and to conform with geographic usage, Palisades dolerite is used in this paper. Locality references are from current U.S. Department of the Interior, Geological Survey, 7.5 minute series (Topographic) quadrangles. Specimens referred to are entered in the rock collection of the Australian National University.

LOCATION AND PHYSIOGRAPHY

The Palisades Sill crops out in New Jersey and New York states (Fig. 1), and forms in part the Palisades along the Hudson River, opposite and north of the City of New York. Its outcrop extends from Staten Island in the south to Haverstraw in the north, then from there west to Mt. Ivy, a distance of about 50 miles, and is up to 1.5 miles wide. The outcrop at Jersey City arises above the low-lying country there and to the south, and from Weehawken north to Haverstraw it forms the western bank of the Hudson River, except in the Nyack area, where it curves inland at Piermont and back to the river bank at Verdrietege Hook. From here it reaches its highest point, 827 feet, at High Tor, and then swings inland to Mt. Ivy, losing its relief, and again passes beneath the surrounding sediments north of Pomona.

Detailed cross sections are shown in Figures 2 to 5, but a general cross section of the intrusion from east to west somewhere between Weehawken and the town of Palisades would show a columnar cliff (Fig. 2) up to 200 feet high, which, with underlying sediments that are mostly concealed beneath talus slopes, reaches, in places, 500 feet above sea level. From the crest of the cliff, the outcrop, which is striated from ice action, slopes gently west to the top contact, though the erosion plane now is incised in places by valleys, many of which formed along faults.

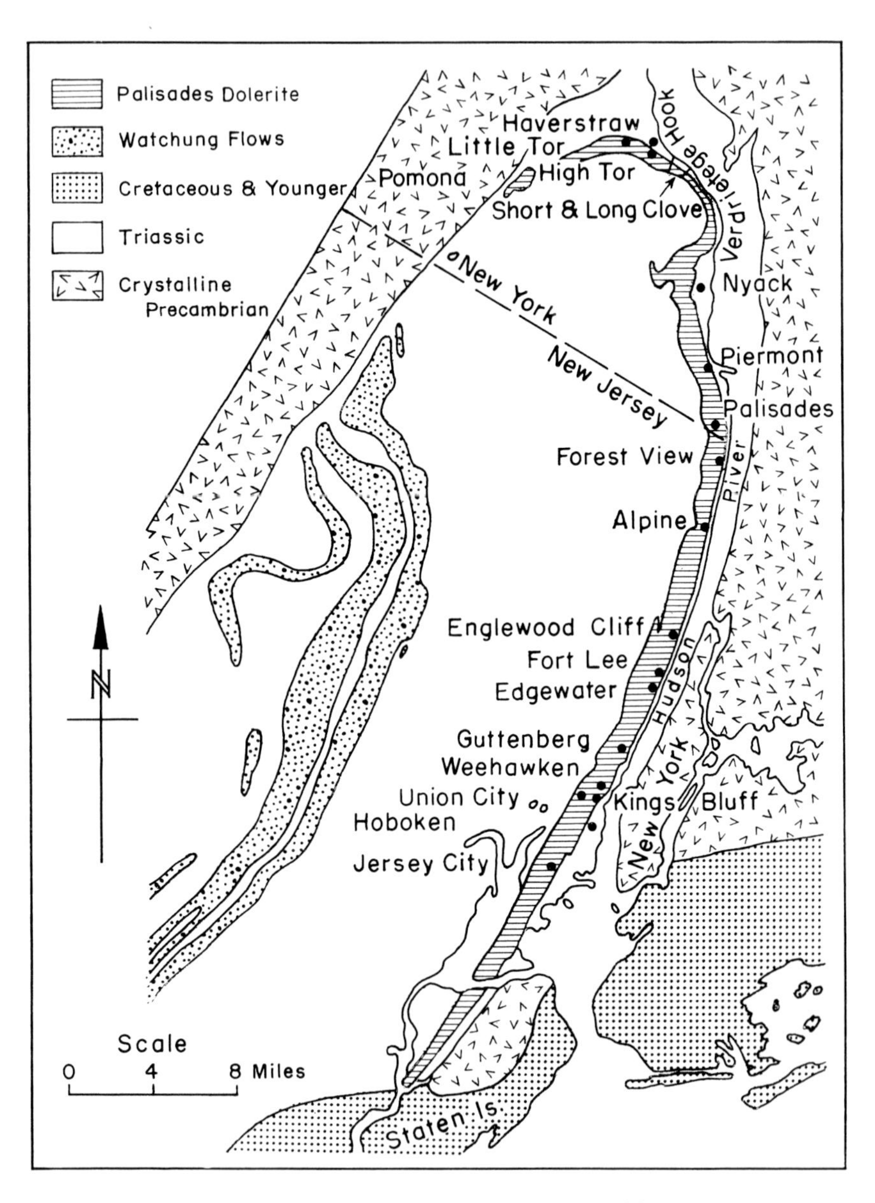

Figure 1. Locality map of the Palisades Sill.

DETERMINATIVE METHODS AND ACCURACY

The experimental methods used for field sampling, sample preparation, mineral separation, petrographic and chemical study of the rocks and minerals are reviewed. All methods proved satisfactory except that some mineral separations by the usual magnetic and heavy liquid procedures did not yield a mineral product entirely representative of the pure mineral phase sought from the rock because microscopic inclusions, including patchy alteration, in some mineral grains modified the physical properties of the mineral sufficiently to bias the separation. This has been shown by chemical work, and the problem is discussed further in the sections on the chemistry and differentiation trends. In addition, the large range in grainsize and the varied texture, including mineral intergrowths, make the separation of pure mineral fractions difficult from a hypabyssal intrusion of tholeiite. Attempts to separate pigeonite from augite were abortive.

Field Sampling

Specimens were accurately located in the field with respect to the upper and lower contacts. Straight-line traverses were made at right angles to the lower contact, and specimen locations were fixed by chain, compass, and Abney level measurements. The locations were checked by aneroid barometer for height above the lower contact; heights were corrected for changes in atmospheric pressure at the lower contact during observations.

Sample Preparation

Type specimens were selected by thin-section examination. About 1 kgm of rock was reduced by rock splitter to 3 cm fragments, from which 400 gm of fresh rock were randomly selected and treated in a small jaw crusher to pass a 5-mesh sieve. The entire product of this crushing was ground in a sintered corundum mechanical mortar grinder until it passed a 120-mesh bolting cloth sieve which was mounted in a perspex frame. Sample loss was avoided during all stages of preparation after the selection of the

3 cm fragments. The homogenized −120-mesh fraction was prepared for the silicate and spectrographic analyses of whole rocks. Rock powders were homogenized by shaking them thoroughly by hand for an hour in a glass bottle; analyses of randomly selected fractions of a sample showed that the powder was homogeneous.

Mineral Separations

Mineral separations were made on the −120- to +180-mesh fraction. This fraction was washed in alcohol to remove rock dust. Microscopic examination showed that the fraction consisted mainly of discrete mineral grains. The usual procedures for the Frantz Isodynamic Magnetic Separator were followed. The magnetic opaque iron mineral was removed by hand magnet from the crushed rock fraction before undertaking the separations. The magnetic separator gave good concentrates of olivines, hypersthenes, monoclinic pyroxenes, and a concentrate of plagioclase with quartz and apatite impurity. Final purification of minerals was achieved with successive heavy liquid separations and centrifugings. For these, a range of bromoform and methylene iodide solutions were used. In a few cases, particularly fayalite, final impurities were removed by hand picking under a binocular microscope. To check the purity of the minerals separated, a few hundred grains were mounted in Canada balsam on a slide with cover slip, and grain counts were made. In all cases mineral concentrates were better than 95 percent pure.

Index of Refraction Measurements and Plagioclase Determinations

The plagioclase in most stages of fractionation is strongly zoned and complexly twinned. Its composition was determined by measuring the index of refraction of the glass obtained by fusion of about 1 gm of mineral powder in a 5-kilowatt radio frequency generator. Provided that the plagioclase separated from the dolerite by magnetic and heavy liquid procedures is representative, an accurate average composition has been obtained. A molybdenum crucible containing the plagioclase charge was suspended by fine tungsten wire in a sealed pyrex tube, in which an inert atmosphere was maintained by a flow of argon. As heating is by induction, a crucible of constant size and shape was used which permitted rapid heating to the melting point of plagioclase. To provide an accurate comparison of compositions, a constant weight of plagioclase was fused at the same temperature for the same time in each case.

As a check on the accuracy of the fusion technique for plagioclase determinations, two analyzed plagioclases (Hess, 1960, Table 10, p. 43, Specimen nos. EB 38 and 7510) were fused under the same experimental condi-

TABLE 1. COMPARISON OF PLAGIOCLASE COMPOSITION BY FUSION TECHNIQUE AND CHEMICAL ANALYSIS

Specimen	Glass RI (25°C)	An Content Fusion Method	An Content Chemical Analysis
EB 38	1.556	An 80.5	An 80.2
7510	1.550	Au 74	An 76

tions. The An content obtained by fusion technique is compared with that obtained by chemical analyses in Table 1.

The results indicate that oxidation and loss of volatiles were insignificant under the conditions of rapid fusion employed. The curve of Schairer and others (1956) was used to relate R.I. to An content. The calcium content of the plagioclases was obtained from a curve relating Ca to An content compiled from data given in Hess (1960, Table 10) for analyzed plagioclases from the Stillwater, Bushveld, Great Dyke, and Skaergaard intrusions.

Refractive indices of plagioclase glass were determined by the usual immersion method on a crushed fraction. A set of calibrated refractive index liquids and a sodium light source were used for the determinations. The index being measured was bracketed between liquids 0.002 apart, and the final determination was made by mixing the liquids until the index was exactly matched. The index of the liquid was then measured with an Abbe refractometer and corrected for temperature to 25°C. Determinations of indices of refraction are considered to be better than ± 0.001.

Optic Angle Measurements

The Universal Stage was used to measure the optic axial angle of the orthopyroxenes and of some olivines. Optic angle measurements were made by direct rotation on a Lietz 5-axis stage from one optic axis to the other, and the usual corrections were applied. A sodium light source was used. Reference was made to Deer and others (1962, v. 1, Fig. 11; 1963, v. 2, Fig. 10) to obtain the composition of these minerals from 2V measurements. With care, angles can be measured to within ± 1°.

Micrometric Analyses

The Swift point counter was used for the micrometric analyses of minerals in type specimens. An area from 10 to 14 sq cm was traversed in thin sections, traverses being 1 mm apart and the interval between points being .33 mm. This gave from 3000 to 4000 counts for each rock, from which a reliable estimate in volume percent of major minerals was made. With the regular subophitic texture (a feature of all rocks except those in the Mg-olivine layer and in the granophyric zone), estimates of the abundance of minor mineral phases, which are interstitial and of small grainsize relative to the major mineral components, are subject to a large error. Modes were calculated from the micrometric determinations by estimating the density of minerals from their composition and converting volume percent to weight percent.

Optical Spectrograph Determinations

The optical emission spectrograph was used for all chemical determinations, except for silicate analyses and Mg and Fe determinations in olivines of the Mg-olivine layer; the results for these are attributed in each case to the analyst concerned.

Three optical spectrographs were used during the course of the investigation, and major and trace element determinations have been made in this way. Analytical work at Yale and at A.N.U. was done on Jarrell-Ash 3.4 m Ebert photographic spectrographs, which have a plane grating with 15,000 lines per inch and 5 Å/mm dispersion. Each instrument is fitted with a Varisource source unit. Plates were read on JAco console comparator microphotometers models 2100 and 2310. The Jarrell-Ash Seidel Calculating Board was used where necessary for low intensity line and background values. In the Bureau of Mineral Resources, a Hilger and Watts 3 m Polychromator was used which has a concave grating with 14,600 lines per inch and 5.8 Å/mm dispersion. The direct reader has a source unit which provides a constant current DC arc from 2 to 15 amps.

Details of analytical procedures are summarized in Table 3. The procedures are based on methods developed by Turekian and by Ahrens and co-workers. Slight modifications in method, particularly in techniques, led to improved analytical precision. All procedures employ a constant current DC arc and anode excitation.

The opaque iron minerals were analyzed satisfactorily by the following optical emission procedure. Two parts of an opaque were mixed with 3 parts of dolerite, W-865-60, and the analyses were done by procedure 7 (Table 3). Element values for the opaques were obtained by subtracting the contribution made by W-865-60. The composition of this rock has been well established by repeated analysis by various methods.

For analytical control, appropriate primary or rock standards were used; in no case were synthetic standards employed. Element values for standards used in working curves are given in Table 2. The standards provided good control for all element determinations except possibly for samples whose Cr, La, Ni, V, and Y values are above 1000 ppm.

The precision for analytical procedures 1 to 4 is discussed in Turekian and others (1957, p. 45 and 46) and in Carr and Turekian (1961, p. 16 and 17). Precision within 12 percent relative deviation can reasonably be expected for most element determinations run in duplicate in the dolerite-granite composition range. For the remaining analytical procedures used, Ahrens and Taylor (1961, p. 189), and Taylor and Kolbe (1964, Table 1) report precisions as relative deviations ranging from 4 to 6 percent for Ba, Ca, Cr, Fe, Ga, La, Mg, Mn, Ni, Ti, and V, to 7 to 10 percent for B, Co, Cu, Nd, Pb, Sc, Sr, Y, and Zr. Precision of analyses by direct reader in duplicate runs using similar procedures to Ahrens and Taylor, has been shown by repeated analyses of W-1 and Sy-1 to be within 5 percent relative deviation for all elements reported.

Because of the repeated analysis of the rocks and minerals used in this study by various optical spectrographs and analytical procedures, I consider that the element values reported are well within 10 percent, and many are probably within 5 percent, of the true values. This statement of accuracy assumes that the element values used for rock standards (Table 2) are accurate; trace values above 1000 ppm are excluded, however, for it is recognized that lack of sufficient natural standards in this concentration range for some elements may have led to insufficient control of working curves.

TABLE 2. ROCK STANDARD VALUES USED IN WORKING CURVES

	Al (%)	Ca (%)	Fe (%)	Mg (%)	Mn (%)	Ti (%)	B (ppm)	Ba (ppm)	Ca (ppm)	Cr (ppm)	Cu (ppm)	Ga (ppm)
NBS99	10.08	0.26	0.05	0.035	0.001	0.01		110				
G-1	7.45	0.97	1.32	0.21	0.02	0.156		1250		19	13	18
GR	7.86	1.85	2.85	1.44	0.043	0.37			11		280	
T-1	8.71	3.70	4.21	1.15	0.077	0.36		650	13	21	48	
Sy-1	4.92	7.22	5.76	2.45	0.31	0.29	70	350	19	54	21	20
W-1	7.90	7.81	7.74	3.93	0.132	0.647	17	170	50	120	110	16
NBS4975	0.20				0.085				145	1800		
NBS4986	3.2			15.55	0.093					1000		
NBS98	13.51	0.15	1.4	0.43	0.004	0.85						
BCS269		0.24		0.56	0.016	0.88						

	La (ppm)	Mo (ppm)	Nb (ppm)	Nd (ppm)	Ni (ppm)	Pb (ppm)	Sc (ppm)	Sr (ppm)	Sn (ppm)	V (ppm)	Y (ppm)	Zr (ppm)
NBS99								120			10	
G-1	100	7		55		49		260		15	15	210
GR	65				51						18	
T-1	22				11			330				175
Sy-1	220		150	350	44	495	14		10	73	450	3000
W-1					78	10	35	180	2.5	240	24	115
NBS4975												
NBS4986					1500			400				
NBS98	80						21			150		
BCS269		330									32	

TABLE 3. DETAILS OF OPTICAL SPECTROGRAPHIC METHODS

Analytical Procedure	1	2	3
Element determinations	Co, Cr, Cu, Ni	Ba, Sr	Ca, Fe, Mg
Mixture	1:1-J.M.90 $CaCO_3$: Sample, mix and grind to −300 mesh in agate mortar	As for 1	2:1 NC Sp-2: Sample, mix and grind to −300 mesh in agate mortar
Amount analyzed (mg)	5	5	5
Arcings rocks— standards—	2 3	2 3	2 3
Electrodes NC preformed Anode Cathode	L4000 L3869 (pointed)	As for 1	L4006 L3869 (pointed)
Wavelength range (Å)	4650-7150 (1st order) 2325-3575 (2nd order)	2600-5100 (1st order)	2000-4500 (1st order)
Electrode gap (mm)	4	4	4
Entrance slit (μ)	10	50	10
Exit slits (μ)			
Entrance slit wedge (mm)	1	1	1
Transmission (%)	50	2	100
Step sector (ratio)			
Filter		Glass	
Stallwood jet			
Amperage (high-low)	16.5 (high)	16.5 (high)	10 (high)
Arcing time (secs)	To completion (75-80)	To completion (75-85)	To completion (about 80)
Pre-burn (secs)			
Emulsion	Kodak SA1	Kodak SA1	Kodak SA2
Developer Kodak D-19 at 20°C (mins)	3	3	3
Fixer Kodak rapid (mins)	3	3	3
Plate calibration	Stepped Fe[x] spectrum[a] (8 step rotating sector 1.6:1)	Stepped Ca* spectrum[a] (8 step rotating sector 1.6:1)	As for 1
Background correction	Yes	Yes	Yes
Analysis line (Å)	Fe[x]3428(x2) Co3453(x2) Fe[x]2840(x2) Cr2843(x2) Fe[x]3266(x2) Cu3274(x2) Fe[x]3428 Ni3414	Ca*4579 Ca*4579 Sr 4607 Ba 4554)	Ca2997, Fe[x]3008, Mg2781
Variable internal standard	Fe	Ca	
Internal standard			
Rock standards (Work curve control)	W-1, G-1, GR NBS4975 and 4984	W-1, G-1, NBS4975	W-1, G-1, NBS4975

(1), (3), (4)	Analytical procedure after Carr and Turekian (1961, p. 12 and 17)
(2)	Analytical procedure after Turekian, Gast, and Kulp (1957, p. 44)
(5), (6), (7)	Analytical procedure after Ahrens and Taylor (1961, p. 189)
(1) to (6)	Analyses by Jarrell-Ash 3.4 meter Ebert Spectrograph
(7)	Analyses by Hilger and Watts 3 meter Polychromator
(a)	Photographic plate calibration curve constructed from stepped analytical line marked [x] and *. Spectrum of rock standards included in each plate to control variations between plates and to compile working curves
(b)	2-step method used for plate calibration. Steps 3 and 4 used for ranges 2772-2840, 3016-3091, 3392-3475 Å (*see* Shaw, 1960). Spectrum of rock standards included in each plate to control variations between plates and to compile working curves

TABLE 3. (CONTINUED)

4	5	6	7
Co, Cu, Mn, Ni, Sc, Ti, Zr, Ca, Fe, Mg	B, Ba, Co, Cr, Cu, La, Nb, Nd, Ni, Sc, Sr, V, Y, Zr	Cu, Ga, Pb, Sn	Al, Ba, Ca, Co, Cr, Cu, Fe, Ga, La, Mg, Mn, Mo, Ni, Sc, Sr, Ti, V, Y, Zr
As for 3	2:1-NC Sp-2 with 0.04% JM592 $(NH_3)_4Pd(NO_3)_2$: Sample, mix in Spex 5000 mill for 10 mins	1:2 Analar K_2SO_4 with 0.1% JM385 In_2O_3: Sample, mix in Spex 5000 mill for 10 mins	As for 5
5	50	100	25
2 3	2 3	2 3	2 3
As for 3	L4261 L3863	L4018 L3863	L4260 L3863
2100-4600 (1st order)	2450-4950 (1st order)	1960-4200 (1st order)	2500-6100 (1st order) 2000-4250 (2nd order)
4	8	8	4
10	40	40	15
			50
1	10	10	
100			100
	Rotating 7 step (2:1)	As for 5	
			Glass for Ba4934, Ca4425, Cr4254, La4333, Mn4783, Sr4607, V4379
		O_2 at 4 litres per min	
16.5 (high)	12.5 (high)	4.5 (low)	8 (low)
To completion (75-90)	To completion (about 140)	To when volatile elements distilled (about 135)	To completion (130)
	7 (Gap at 0.25 mm)	2 (reduced gap)	2 (reduced gap)
Ilford N50	Ilford N30	Kodak 103-0	
5	5	4	
5	5	4	
Stepped Fe spectrum[b] (7 step rotating sector 2:1)	Self calibration method[c]	Self calibration method	
Negligible	Yes and Seidel function where necessary	Seidel function scale	No
Ca3006, Co3453, Cu3274, Fe2929, 2937, Mg2780, 2781, 2938, Mn 2801, Ni3414, Sc4246, Ti2956, 3242, Zr3438	B2498, Ba4934, 4554, Co3453, Cr4254, Cu3274, La4333, Nb3163, Nd4303, Ni3414, Sc4246, Sr4607, V4379, Y3327, Zr3438	Cu3274, Ga4172, Pb3683, 4057, Sn3175	Al2652(x2), Ba4934, Ca4425, Cr4254, Co3453, Cu3274, Fe3008, Ga2943, La4333, Mg2780, Mn4783, Mo3170, Ni3414, Sc4246(x2), Sr4706, Ti3990, V4379, Y3327, Zr3438(x2)
	Pd(3421Å)	In (3256Å)	Pd (3421Å)
W-1, G-1, Sy-1 NBS4975 and 98	W-1, G-1, Sy-1 NBS98 and 99	W-1, G-1, Sy-1	W-1, G-1, GR, Sy-1, T-1, BCS269, NBS4975, 4986, 99

(c) Ahrens and Taylor (1961, p. 159-161). Spectrum of rock standards included in each batch of plates of analytical run to control variations in plates between runs and to compile working curves

JM Johnson Matthey

BCS British Chemical Standards

NBS U.S. National Bureau of Standards

NC National Carbon

(x2) Second order

Electron Probe Determinations

An Electron Probe X-ray Microanalyser (ARL model EMX) was used for the quantitative determinations of Fe and Mg in olivines of the Mg-olivine layer. The determinations were made simultaneously using K α radiation. The electron beam, which is 2 to 3 microns in diameter, was focused to a spot in a polished thin section of the rock whose surface was thinly coated with carbon. The accelerating voltage was 12 KV and the specimen current 0.04 microamps. Spot analyses were made at intervals of 1 to 10 microns with an integration time of 70 seconds. Corrections were made for beam current fluctuations and for background, but fluorescence, matrix absorption, and atomic number affects were minimized by using calibration curves controlled by standard olivines (Suzimaki, Marjalahti, and so on). Determinations are considered accurate to $\pm$ 0.2 percent for Fe and 0.5 percent for Mg, this estimate being derived from the reproducibility of results on standards, and from the linearity of the calibration curve.

GENERAL GEOLOGY

The Palisades Sill is a unit in the series of basalt flows and sills emplaced in a structural and stratigraphic basin called a taphrogeosyncline by Kaye (1951); though no longer continuous along the strike, the geosyncline originally extended from North Carolina to Nova Scotia along the eastern coastal strip of the United States. It is the intrusive equivalent of the Upper Triassic Watchung basalt flows farther west, and its age has been determined by Erickson and Kulp (1961) at 190 ± 5 m.y. by a K-Ar determination on biotite from a dolerite at Fort Lee. It intruded the Newark Formation, which comprises mainly arkoses, shales, and sandstones, and its thickness appears to range from 750 to 1200 feet, though probably it is mostly between 1000 and 1100 feet. Where contacts are concordant, as between Weehawken and Nyack, they show that the intrusion has a general strike of N. 30° E., and dips between 10° and 15° WNW.

Apophyses to the main intrusion include small sills in strata immediately above and below it (Pl. 2), and satellite intrusions above it, such as the small laccolith at Granton (*see also* Kummel, 1897).

Though long referred to as the Palisades Sill, and the name is retained in this paper, it is probably in part a sheet and in part dike-like. Only recently Thompson (1959) and Lowe (1959) discussed its form. Thompson believes it is a sill throughout, whereas Lowe supports the interpretation originally advanced by Darton (1890) and Kummel (1900) that the discordant contact relations revealed in places indicate its dike-like character in part, and do not result exclusively from faulting as maintained by Thompson.

Complete understanding of its form will require more detailed field work supported by geophysical work. My field and laboratory observations still leave the problem open, but go some way toward confirming that the intrusion was emplaced as a sheet, but that it has some dike-like features in part, despite the presence of faults in places. There seems little doubt that the intrusion is in discordant contact with the enclosing sediments, both above and below, at a number of places. The significance of this and of faulting is discussed below.

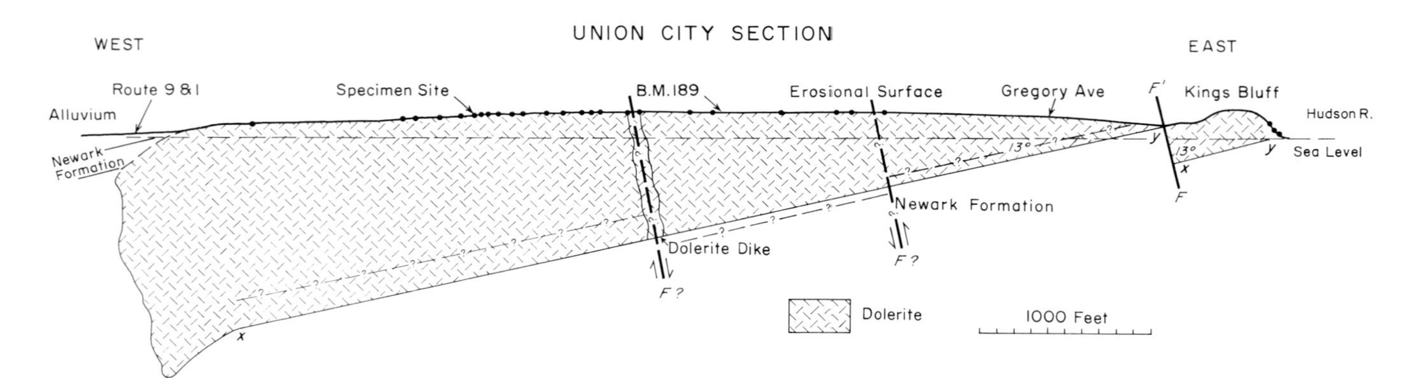

Figure 2. Cross section of the Palisades Sill. Traverse line is 300 feet south of bench mark (B.M.) 189 and is along the approach to the Lincoln Tunnel.

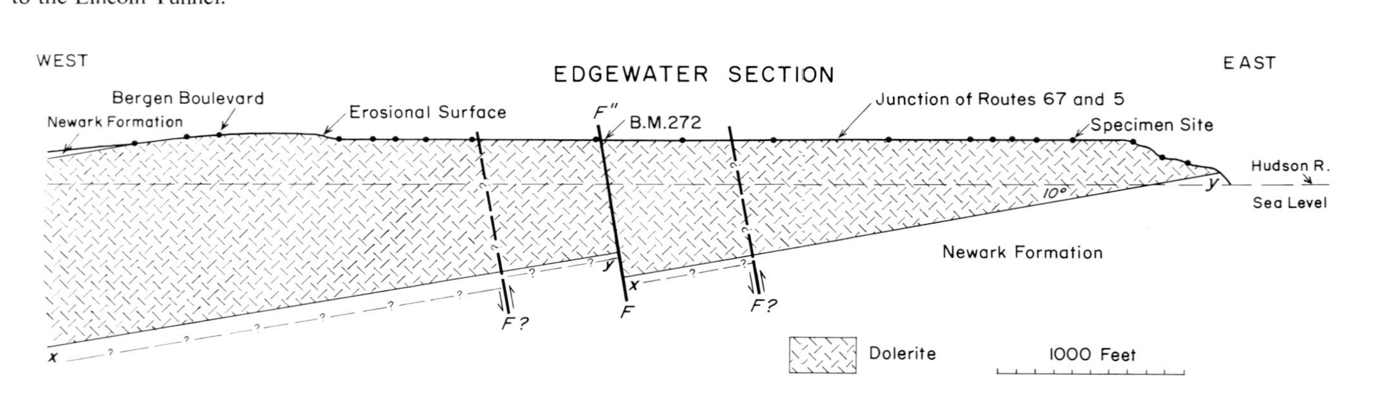

Figure 3. Cross section of the Palisades Sill. Traverse line intersects bench mark 272.

FIELD OBSERVATIONS

Contacts

The basal contact is mainly concordant with the sediments where observed between Hoboken and Verdrietege Hook. However, breaks occur where the intrusion cuts across from one sedimentary horizon to another, or where faults dislocate it. Such displacements range from a couple of feet to 100 feet or more, and offsets up to 200 feet can be seen in a number of places, for example, Fort Lee, Alpine, Piermont, and Verdrietege Hook (*see also* Darton, 1890; Kummel, 1900, for examples). The basal contact ranges from a few feet above sea level, as at Kings Bluff, to 300 feet above at Forest View, and climbs steadily from south to north, broken in places by displacements. From Short Clove to Mt. Ivy, much of the contact is reported to be 300 to 500 feet above sea level and at an extremely low angle to the sediments (Lowe, 1959, p. 1133), or is concordant where observed (Thompson, 1959 p. 1115-1117).

In the field the top contact appears to be concordant between Guttenberg and Nyack, and roughly parallels the Hudson River about 1.25 miles to the west. South of Guttenberg the contact is discordant in places, as for example in the western portal of the New York Central West Shore Tunnel at Weehawken, where a steeply dipping contact of chilled dolerite abuts against gently dipping arkosic and quartzitic sediments. Again, between Long Clove and Mt. Ivy, the southern contact, where observed along South Mountain Road, appears discordant, and at the southern portal to the New York Central West Shore Tunnel near Long Clove the chilled dolerite cuts abruptly across slightly hornfelsed sediments that dip gently north.

In the Englewood Cliff Section, the type section, the top contact relations were complicated by faulting and glaciation. About 50 feet is missing from the top of the intrusion at Old Coytesville Quarry. A top chilled contact occurs about 2000 feet west of the quarry in Flat Rock Brook, off Route 4, but this is isolated from the Englewood Cliff Section by a major fault (Fig. 4, FF″). At the Old Coytesville Quarry the contact relations are deceptive; glacial abrasion removed the top chilled contact and exposed an early dolerite stage. During the course of the ice action, large slabs of bedded

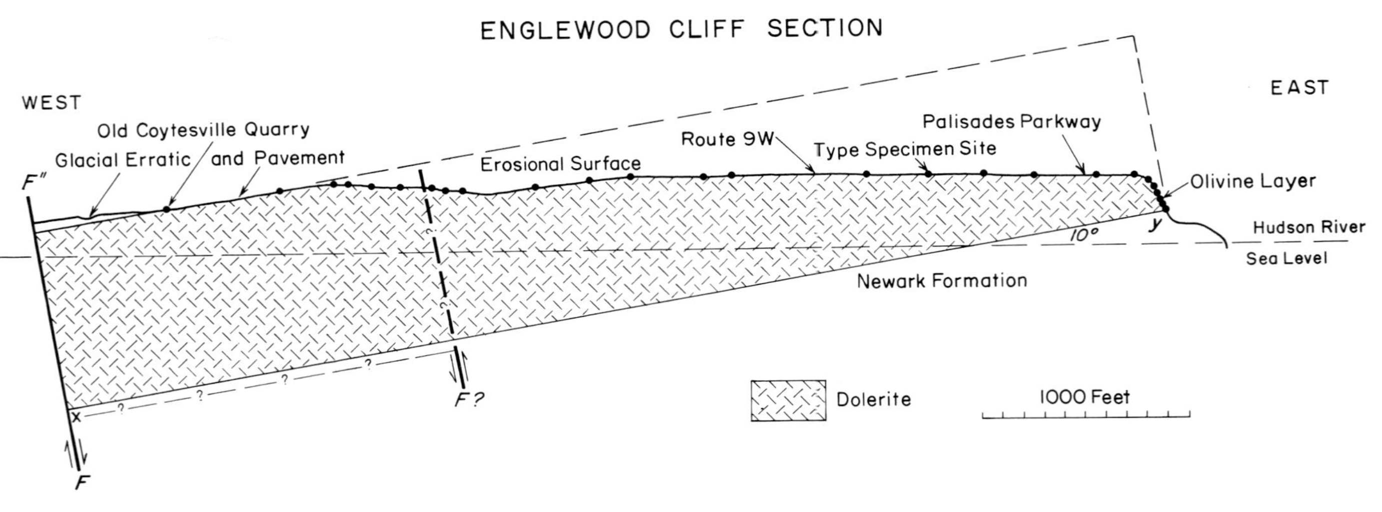

Figure 4. Cross section of the Palisades Sill. The traverse line intersection with Route 9W is 700 feet south of bench mark 371.

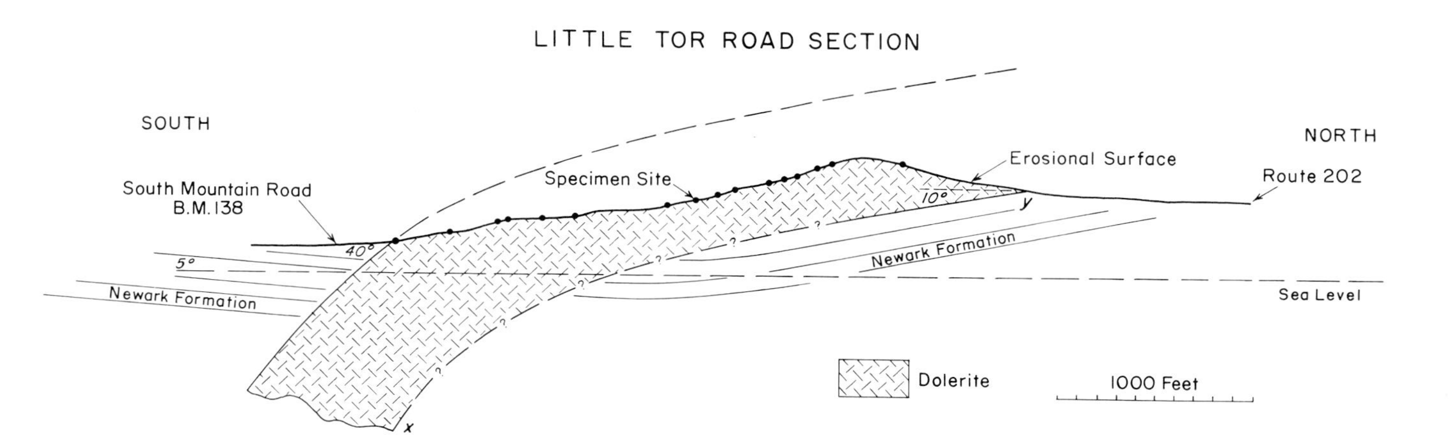

Figure 5. Cross section of the Palisades Sill. The traverse line is along a road being formed between bench mark 138 and Route 202. Numerous minor faults not shown.

Triassic sediments were moved into a conformable resting position on the striated pavement, creating a relationship between dolerite and overlying sediment reminiscent of a concordant igneous contact.

The sediments at some discordant contacts are distorted, suggesting movement during magma consolidation or during subsequent tectonic activity that faulted the rocks of the area.

Faults and Joints

Faults are fairly common, particularly minor breaks; however, it is the major breaks that complicate the petrological analysis of the intrusion, and their identification and the estimation of their displacements are difficult in field exposures. Petrological evidence has been used to assess the effect of some faults on the results of the investigation.

The sections (Figs. 2, 3, and 4) show the location of established faults according to the State of New Jersey Geological Map (Scale 1:250,000, Atlas No. 40), compiled by the Department of Conservation and Economic Development. Other faults have been deduced from petrological evidence and are shown as inferred faults in the sections (Figs. 2, 3, and 4), located where discontinuities are indicated in the sequence of fractionation established for the intrusion. Alternative explanations for the discontinuities were considered unlikely; for example, that the breaks in the sequence occur at internal intrusive contacts.

Sections measured south of Alpine show that the thicknesses of corresponding mineral horizons representing the various stages in fractionation are fairly constant in each, and that they are continuous along the length of this part of the intrusion. The Englewood Cliff Section is complete except possibly for some of the fayalite granophyre horizon and the top chilled dolerite. The Union City Section shows that the fayalite granophyre horizon is about 100 feet thick, and as most sections examined in the southern half of the intrusion contain the ferrohypersthene and granophyric dolerite horizons, it is possible that part of the fayalite granophyre horizon has been faulted out of the Edgewater and Englewood Cliff sections. Other inferred faults shown in the sections were located on a similar basis. The location of the faults deduced in this way appears logical in terms of geological information portrayed in published maps of the area, which implies that the southern part of the intrusion is transgressed by a series of *en echelon* faults with an approximate north-northeast trend.

The sub-surface attitude of faults is not known, and for this reason they have been shown perpendicular to the base of the intrusion. Normal faults were probably concomitant with magma injection and with contraction of

COLUMNAR CLIFF OF BASAL PART OF SILL,
GEORGE WASHINGTON BRIDGE

Columnar cliff of the Palisades Sill adjacent to the George Washington Bridge.

BASAL CONTACT AND Mg-OLIVINE LAYER, KINGS BLUFF

Kings Bluff exposure showing the concordant basal contact on the left and a minor transgression across the Newark sediments on the right. At this locality the Mg-olivine layer is the narrow zone in the top half of the plate which roughly parallels the transgressive basal contact. It occurs immediately above the chilled basal dolerite. The person is pointing to an apophysy of dolerite that has wedged between hornfelsed strata below the main intrusion.

the intrusion upon solidification. Possibly the late chilled dolerite dike (Fig. 2) exploited a fault plan for injection, which suggests that epeirogenic movements continued until basic igneous activity ceased in the region. The *en echelon* faults referred to above, however, are probably due to compressional deformation which accompanied folding of the Triassic sediments.

Columnar joints are conspicuous in the cliff (Pl. 1) and quarry faces, and in exposed surfaces west of the cliff, where sections of columns can be seen in many places. The columns pass without deflection through the Mg-olivine layer. In the cliff face, and in exposures near the top contact, a series of joint fractures developed parallel to the contacts.

Intrusive Form and Thickness

The evidence from contacts suggests that the intrusion is either a sheet that has some pronounced discordances because of the transgression of magma across strata at various places in the top and bottom contacts, or that it is in part a sheet and in part dike-like.

South of Palisades the intrusion is a gently dipping essentially concordant tabular body. This is shown by exposures of the basal contact, and by the orientation of columns in columnar jointing of the Englewood Cliff Section. The joints form the columns in the escarpment and crop out at many places along the back slope toward the top contact.

In this part of the intrusion the basal contact dips from north to south, and hence the present erosional surface in the north exposes a deeper section than that in the south. Moreover, the cumulative affect of transverse faults possibly has contributed to the elevation of the northern part of the intrusion. Indeed, the Triassic sediments in the northern section have a distinct axial plunge to the south-southwest, and exposures of sediments along South Mountain Road contain horizons of conglomerate and grit which are probably basal members of the Newark Formation. The structure of this sector would be controlled largely by the behavior and proximity of the crystalline basement.

The most attractive interpretation on present field and laboratory evidence is that the intrusion is a sheet which has some dike-like characteristics, particularly between Short Clove and Mt. Ivy. The interpretation of the form and attitude of the intrusion in the Little Tor Road Section (Fig. 5) was influenced by diamond drill hole data obtained at Haverstraw Quarry (Lowe, 1959), as well as by the further field and petrological observations made in this investigation. Available evidence is still insufficient, so that the interpretation remains conjectural.

Possibly the semicircular northern outcrop of the Palisades steepens with depth into a cone sheet, the western half of which is still concealed beneath Triassic sediments. Jurassic dolerite cone sheets are not uncommon in Tasmania, where, it is believed, they serve as feeders to many intrusive sheets, and this may also be true for the Palisadan province.

It cannot be ruled out that the Palisades intrusion south of Short Clove becomes more dike-like as it extends westward into the subsurface. At Byrum, on the Delaware River, some contemporaneous intrusions have concordant basal contacts and discordant top ones, the discordance becoming more pronounced as the contact steepens with depth. The Palisades Sill may have a similar form.

Thickness can be confidently estimated only for the southern sections examined. Lowe (1959) estimated the thickness at Haverstraw Quarry to be 730 feet. The attitude of the southern half of the sill has been established mainly by projection, using measurements on the outcrop of the basal contact. Some control was gained in the Union City Section by a reported (Lewis, 1908a) basal contact 50 feet below sea level, located in a hole at Gregory Avenue. The basal contact was encountered also in holes evaluating the foundations for the George Washington Bridge. These indicated dips from 10° to 15° WNW.

Disregarding the effects of possible faults and glaciation on the Englewood Cliff Section, the section is about 900 feet thick. To this can possibly be added 50 feet of missing fayalite granophyre and 50 feet of glacial abraded top of the intrusion, making the thickness of the intrusion about 1000 feet at this location. On a similar basis, the readjusted thickness of the Edgewater and Union City sections becomes 1100 and 1050 feet respectively.

The cited height of specimens from the base assumes a concordant intrusion as shown in Figures 2, 3, and 4, where *xy* marks in each case the basal contact, and measured heights of specimens are adjusted only for established faults. Also shown by the queried broken lines is the inferred base after allowing for displacements indicated by the inferred faults.

Mg-Olivine Layer

The famous layer of hyalosiderite dolerite can be seen from Paterson Plank Road, .25 mile south of Riverview Park, Hoboken, north to Nyack Beach State Park. In general, the base of the layer is from 30 to 40 feet above the chilled dolerite which itself can be up to 30 feet thick, though there are definite exceptions to this generalization. The layer appears to be thickest in the Edgewater-Englewood Cliff area, and is conspicuous in road cuttings. At Edgewater it forms a prominent, but somewhat uneven, ledge by differential weathering in the cliff face. Here it can be seen that it migrates up and down 20 feet or so, and that it bulges and pinches along its length, ranging in thickness from 3 to 30 feet, but probably averaging about 20 feet. At

Englewood Cliff, in the type section, the layer is 50 feet from the base of the intrusion and extends to 80 feet above. At Kings Bluff (Pl. 2) it is narrower, 2 to 10 feet thick, and rests directly on chilled dolerite from 25 to 35 feet above the basal contact. The layer also varies in thickness along its length at this locality, and here the basal contact is discordant in part, so that the top of the chilled dolerite, and in turn the olivine layer, roughly reflect the shape of this contact.

From Alpine to Upper Nyack the layer is only a few feet thick in many places. At Forest View it measures 12 feet and is 60 feet from the basal contact, whereas in the old quarry face at Nyack Beach State Park it is about 2 feet thick and 40 feet from the contact. The weathered chilled dolerite reported by Walker (1940, p. 1065) to have a field appearance reminiscent of the olivine layer forms a ledge beneath the layer at this locality. The olivine layer is absent from the Haverstraw Quarry area and from outcrops to the west of there.

Schlieren

Where observed, schlieren of pegmatitic dolerite occur in the top third of the intrusion and measure from a few inches to 4 feet across. They are particularly well exposed in road cuttings through the top contact at Englewood Golf Course, in the ferrohypersthene dolerite horizon in the Lincoln Tunnel approach road, and in the Old Coytesville Quarry. Walker (1953, p. 1065) also records schlieren near the Mg-olivine layer.

Products of the Late Hydrothermal Stage

In the top quarter of the intrusion, products of hydrothermal activity are widespread. Zones of hydrothermally altered dolerite up to 20 feet wide occur in places and can be seen cutting across the fayalite granophyre layer in the Union City Section. Sinuous fractures are common in the chilled marginal zones and have had their walls sericitized and uralitized to form veins up to .5 inch wide. Calcite veins and a wide variety of minerals (Mason, 1960) are products of late magmatic emanations which were emplaced mainly in the upper levels of the intrusion, whereas residual veins of quartz and albite with some analcite up to 1.5 inches wide, are found in both the upper and lower parts.

Assimilation

In outcrops at McKinley School near Columbia Park (Pl. 3), and also at the back of Wards Trucking Yard, Bergen Turnpike near Machpelah Cemetery, partly assimilated sedimentary xenoliths up to a foot in diameter are abundant in the top chilled contact zone. Where assimilation is nearly complete, the contaminated dolerite forms leucocratic patches in the normal dolerite. Assimilation appears to be localized in some of those places where the top contact is discordant, as is the case at Ward's Trucking Yard.

Contamination appears to have been marginal, and did not influence the over-all composition of the magma. Xenoliths of sediment plucked from the floor of the intrusion, which measure from a few inches to many feet across, rest in or near the basal chilled zone, and though contact metamorphosed, have sharp margins and show no signs that the magma has been aggressive toward them. However, as seen about 400 yards south of the George Washington Bridge on Henry Hudson Drive, the basal contact sediments in places have been plastically deformed into tight folds.

Rheomorphic Veins

Rheomorphic veins and dikes up to 15 inches wide are common in the top chilled dolerite zone, the mobilized sediment having penetrated up to 60 feet into the intrusion. Some veins, only an inch wide, penetrate 30 feet. These are well displayed in the Old Coytesville Quarry, Englewood (Pl. 4), and in outcrops in the grounds of the McKinley School near Columbia Park. Poldervaart (1963, written commun.) recognized two types of rheomorphic veins, resulting from (a) fluidization, and (b) partial melting of the overlying sediments.

Metamorphism

Metamorphism of adjacent sediments is limited mainly to hornfelsing, which occurred up to 30 feet from the contacts, but is more pronounced at the lower than the upper contact. Both Kummel (1897) and Lewis (1908a) have described the metamorphism in general terms. In addition to metamorphism, magmatic emanations contributed to the alteration of the top contact sediments (Lewis, 1915). A. Poldervaart (1963, written commun.) observed, at Englewood, tourmaline porphyroblasts in meta-argillites and analcite in meta-arkose; these formed in a zone between plastically deformed sediment at the dolerite contact and sediment farther out whose joints contain radiating hematite.

Late Intrusions

Late basaltic veins transgress the intrusion near the base, below and in the Mg-olivine layer (Lewis 1908a, p. 126, and Walker 1940, p. 1086-1087). Some late chilled dolerite dikes cut right across the intrusion. At Union City, in the cutting that forms the approach to the Lincoln Tunnel, chilled bronzite dolerite dikes, up to 60 feet wide, cut the ferrohypersthenen and pigeonite dolerite horizons in the intrusion.

PARTLY ASSIMILATED XENOLITHS, UNION CITY

Partly assimilated sedimentary xenoliths in the top contact dolerite next to the outcrops in the McKinley School grounds.

RHEOMORPHIC VEINS, ENGLEWOOD

Rheomorphic vein in the top of the Palisades Sill at Old Coytesville Quarry, Englewood.

PETROGRAPHY

Rocks representing the various stages in fractionation of the magma came mainly from the Englewood Cliff-Old Coytesville Quarry Section, which has been chosen for detailed mineralogical and chemical study. Other sections to which reference is made for supporting evidence, or to establish the extent of mineral horizons within the intrusion, are (a) Union City, (b) Edgewater, (c) Haverstraw Quarry diamond drill core numbers 14 and 19, and (d) Little Tor Road. Figures 2, 3, 4, and 5 show the sections and specimen localities. The location of the diamond drill holes in Haverstraw Quarry is given by Lowe (1959, p. 1133) in a section showing holes 15, 18, and 19; hole 14 was sited 530 feet southeast of a point 200 feet north of hole 15. The much-studied Edgewater Section (Walker, 1940) is petrologically similar to the Englewood Cliff Section.

Table 4 lists the detailed locations of the type specimens. The name used for each has been selected on the basis of petrographical type (a description of each follows), and of the intrusive environment and the stage the rock represents in the fractionated dolerite series. The rocks are all variants of tholeiitic dolerite.

To assist in the discussion of the results, some stages have been subdivided, as for example 3(i), where pigeonite crystallized as a primary product. Stages 3(i) and 3(ii) are otherwise considered together, as they are very similar in other respects. Furthermore, until the study of the monoclinic pyroxenes has been completed, the fractionation range shown for some stages in the Englewood Cliff Section must be regarded as approximate only, and the present division between the early and late pigeonite stages, for example, is an arbitrary one.

Description of Type Specimens

Chilled dolerite. The composition of the lower chilled dolerite is fairly uniform along the whole length of the intrusion, both mineralogically and chemically (H. H. Hess, 1956, and 1961, oral commun.), but that of the top contact varies somewhat owing to late-stage hydrothermal activity and to the introduction of K in particular.

TABLE 4. TYPE SPECIMEN DETAILS

Specimen	Section	Height Above Base (ft)*	Equivalent Height Englewood Section (ft)	Fractionation Stage	
W-889LC-60	Englewood Cliff	1			
W-FUC-60	″	900			
W-865-60	″	30		Early Fractionation Stages	1
					2 (i)
W-824-60	″	70			2 (ii)
					2 (iii)
W-804-60	″	90			3 (i)
					3 (ii)
W-U-60	″	215			4
W-R-60	″	365		Middle Fractionation Stages	5 (i)
W-N-60	″	560			5 (ii)
W-J-60	″	685		Late Fractionation Stages	6
W-WU3-61	Union City	990	785		7 (i)
16441	″	≈980	775		
W-F-60	Englewood Cliff	790			7 (ii)
W-E-60	″	805			8
W-D-60	″	830			
W-WU10b-61	Union City	870	675		
14-104 14-105	Haverstraw	103			5 (i)
14-109½ 14-112	″	109			3 (i)
W-WU17b-61	Union City				

*Height of specimens measured from base of intrusion shown *xy* in sections (Figs. 2, 3, 4)

TABLE 4. (CONTINUED)

Fractionation Range Englewood Section (ft)	Rock Name	Factors Determining Divisions
0-20	Chilled dolerite	
20-37	Early dolerite	With primary pigeonite
52-80	Hyalosiderite dolerite	Mg-olivine layer Base Middle Top
85-110	Bronzite dolerite	(i) With primary pigeonite
110-175	Bronzite dolerite	(ii) Free of pigeonite—minor amount exsolution in orthopyroxene
175-310	Hypersthene dolerite	Much exsolution in orthopyroxene 50% inverted pigeonite
310-	Early pigeonite dolerite	(i) Properties of pigeonite consistent with Mg-rich varieties
-640	Late pigeonite dolerite	(ii) Properties of pigeonite consistent with Fe-rich varieties
640-740?	Ferrohypersthene dolerite	
	Fayalite granophyre	Fe-olivine horizon
790-800	Granophyric dolerite	
800-850	Ferrodolerite	
850-900		Sequence reversal toward top contact
	Pegmatitic dolerite	Schlieren
	Pigeonite dolerite	
	Bronzite dolerite	
	Late chilled dolerite	Dike intruded into Palisades Sill

TABLE 5. MICROMETRIC ANALYSES, ENGLEWOOD CLIFF SECTION
In volume percent (weight percent in parenthesis)

Specimen	Height*	Plagioclase	Mg-Olivine	Fe-Olivine	Auguite	Pigeonite	Ortho-proxene	Mauve-brown Ferro-augite
		[0.1]	[1.5]		[1.0]		[0.5]	
W-889LC-60	1	39				54		
W-865-60	30	46 (40.3)	1 P		42.5 (46.9)		4 (4.3)	
W-824-60	70	30 (25.2)	22 (24)		16 (16.9)		22 (23.6)	
W-804-60	90	42.5 (36.8)			24.5 (26.7)	3 (3.2)	23.5 (26.4)	
W-U-60	215	49 (43)			25.5 (28.1)		21.5 (24.4)	
W-R-60	365	58.5 (52.9)			25 (28.5)	8 (9.1)		
W-N-60	560	66 (60.6)			17 (19.7)	7.5 (8.6)		
W-J-60	685	46.5 (40.6)					3.3 (4)	10 (11.2)
W-E-60	805	35 (30.5)		2 ip (2.8)				18 (20.4)
W-F-60	790	40		1.5				

P Pseudomorphs (Vol. %)
ip Including pseudomorphs
tr Trace

The dolerite has been chilled to an extremely fine-grained uniform rock with sharp, and in some cases, welded contacts with the enclosing sediments. In the northern part of the intrusion, some turbid glassy mesostasis, with needles of opaque mineral, is present between the crystalline phases. South of Nyack, the basal chilled dolerite is holocrystalline and intergranular; its main constituents are augite, pigeonite, and plagioclase; olivine, hypersthene, and biotite are much less abundant.

Microphenocrysts form up to 3 percent by volume of the rock, of which up to 1.5 percent can be olivine. They are smaller and less common in the top chilled contact dolerite. Olivine occurs only as microphenocrysts and is invariably corroded, and partly or completely altered, to a serpentine-like

TABLE 5. (CONTINUED)

Pale-green Ferro-augite	Horn-blende	Biotite	Chlor-ite	Sphene	Apa-tite	Opaque Iron Minerals	Quartz and Micro-pegmatite	Alteration Products
						5.0		0.5
		0.5	2		tr	2.5 (4.0)	1.5	
		3		tr	tr	2 (3.0)		5
		0.6	1.5		tr	1.5 (2.4)	2.5	0.4
		0.6	0.3		tr	1.3 (2.0)	1.4	0.4
	1.8	0.6	1		0.1	2 (3.4)	3	
	1	0.5	1	tr	0.2	3 (5.0)	3.8	
21 (23.4)	3	0.2	0.3	tr	1	5 (8.0)	10	
5 (5.7)	6.5	4		tr	0.3	6.2 (10.0)	23	
tr	12.5	15		tr	1.5	3	24.5	2

* Height above base in feet
[ ] Microphenocrysts (Vol. %)

mineral and iron oxide (and to "iddingsite" in the top contact dolerite), which in turn can be surrounded by a reaction rim of orthopyroxene. The olivine is unzoned and its composition ranges from $Fo_{81.5}$ to Fo_{79} at Englewood Cliff.

Most of the pyroxene microphenocrysts are augite with slightly resorbed grain margins, though a few are euhedral or subhedral. Many have undulose extinction, are strongly cleaved, and appear strained; some of the bladed crystals are bent. The microphenocrysts commonly are in glomeroporphyritic aggregates that may also include an occasional bronzite grain; orthopyroxene otherwise occurs as relict cores in a few of the augite grains. The aggregates measure up to 2 mm across, but subindividuals are mostly less than 1 mm; a few of the bladed crystals reach 3 mm. In the groundmass the pyroxene is anhedral and contains quite a noticeable amount of pigeonite, though it would be difficult to estimate the proportion relative to augite. The monoclinic pyroxene grains average about 0.05 mm diameter. The thin laths of plagioclase range from 0.02 to 0.2 mm in length, with a few up to 0.5 mm. It has an average composition of An_{66}.

The orthopyroxene microphenocrysts in the pyroxene glomeroporphyritic aggregates are of similar size but are much less abundant than augite; they show typical patchy extinction and pale-pink pleochroism. At Kings Bluff there is another generation of orthopyroxene which is of late formation; this contains exsolution lamellae, particularly in mantles around relict olivine grains. Many of the grains are incomplete and incorporate groundmass grains. They are identified by their optical continuity, straight extinction, and exsolution lamellae, and many have β perpendicular to the plane of the thin section. Plagioclase laths in the same rock show a rough flow orientation. Indeed, some of the chilled dolerite at Kings Bluff, upon which the Mf-olivine layer directly rests, appears distinct from that at the basal contact, in that it shows the above flow features, is richer in olivine, and contains glomeroporphyritic aggregates of euhedral plagioclase microphenocrysts, but fewer augite aggregates.

Finely divided opaque iron mineral is evenly distributed throughout the basal chilled dolerite. Small flakes of red-brown biotite are more common in the upper than in the lower chilled dolerite. The biotite and quartz, and the patchy uralitization of pyroxene and sericitization of plagioclase, resulted mainly from hydrothermal activity during the final stages of consolidation of the magma. The pleochroic haloes in some of the biotite suggest the presence of minute zircon inclusions.

Early dolerite (1). In the type section, normal dolerite, with a typical intergranular to subophitic texture and fine grainsize, was developed 30 feet from the basal contact. It is taken to represent stage 1 in fractionation of the intrusion. Elsewhere, chilled dolerite extends from 10 to 30 feet from the contact, and above it the grainsize progressively increases to the Mg-olivine layer.

The early dolerite contains glomeroporphyritic aggregates of anhedral augite or augite and bronzite grains up to 3 mm across; they contain subindividuals from 1 to 2.5 mm diameter and may include blades of augite up to 2 mm by 0.4 mm. These are much more abundant than the corresponding aggregates in the chilled dolerite below, but are similar in their textural and mineralogical features. The crystalline base consists of plagioclase, augite, and pigeonite. A visual estimate indicates that about a third of the monoclinic pyroxene is pigeonite. Minor constituents are opaque iron mineral, biotite, micropegmatite, and apatite.

The plagioclase laths in the groundmass range from 0.1 to 0.8 mm in length and have an average composition of $An_{58.5}$. A few small, much more basic, euhedral laths are enclosed in some of the pyroxene grains.

Spectrochemical analysis of the augite indicates that it is a normal variety. A few augite crystals contain cores of bronzite. The anhedral and subhedral grains in the groundmass range from 0.05 to 0.2 mm across. The composition of the pigeonite has not been determined, though numerous grains of primary crystallization can be readily identified by their near-uniaxial positive interference figures.

The bronzite microphenocrysts show typical pink pleochroism and patchy extinction, and lack zoning or exsolution of clinopyroxene, though in a few cases augite built on and enveloped them. The orthopyroxene may also occur as large grains in the glomeroporphyritic aggregates. It has an average composition of En_{83}.

The few corroded olivine grains present measure up to 1 mm across. In glomeroporphyritic aggregates of pyroxene, and elsewhere, there are patches of a serpentine-like mineral and iron oxide after olivine. A few pseudomorphs of yellow "iddingsite" after olivine are also present.

The opaque iron mineral formed small anhedral grains and rods throughout. Small amounts of micropegmatite are associated exclusively with the plagioclase areas of crystallization.

Hyalosiderite dolerite (2). The distinctive Mg-olivine layer, which extends from 50 to 80 feet above the basal contact in the type section, is a well-known feature of the Palisades Sill. Its texture is gabbroic and poikilitic; and the essential minerals have a fairly equant shape. The texture, and the abrupt disappearance of olivine at the top and the rapid increase in olivine content at the base of the layer, sharply delineate it from the subophitic dolerites above and below. The dolerite in the layer consists of plagioclase, hypersthene, olivine, and augite, and contains subordinate biotite, ilmenite, titanomagnetite, and rare chalcopyrite and apatite.

The plagioclase grains are subhedral and equant, and range from 1 to 2 mm across. In the middle of the layer they have an average composition of $An_{63.5}$. Toward the base and top of the layer numerous small laths of plagioclase are included in the pyroxene, and, in particular, in the orthopyroxene; the plagioclase is also subophitically intergrown with them. In the center of the layer these inclusions are rare; here nearly all the plagioclase formed large subhedral grains in which olivine inclusions are abundant. The plagioclase and olivine inclusions impart a poikilitic texture to the dolerite.

Olivine occurs as large grains ranging from 0.4 to 1 mm diameter, with a few to 2 mm, and as small grains from 0.02 to 0.25 mm across (most are less than 0.1 mm). The large grains are commonly corroded, and some contain irregular cracks along which serpentine-like and opaque iron minerals developed. Many occur in areas of pyroxene crystallization where they have anhedral boundaries with pyroxene and subhedral or euhedral with plagioclase. The smaller grains are euhedral and subhedral, and mostly occur in plagioclase areas of crystallization (Pl. 5); some are also in augite, and less commonly in orthopyroxene where their grain margins are rounded and suggest resorption. Microprobe analysis shows that the distinction between the two groups is apparent rather than real. Grains are zoned, and the largest are generally somewhat poorer in the forsterite molecule, contrary to previous belief (Walker, 1940 p. 1068). Zoning is less marked in the small grains because of their size, but where they occur within a large zoned plagioclase crystal, they are progressively more Fe-rich from center to margin of the plagioclase. The range of composition varies consistently from the bottom to the top of the layer (*see* Table 6 and Fig. 7), olivine being on an average most Mg-rich in the center of the layer, where the range in composition is Fo_{77} to Fo_{69}.

Orthopyroxene straddles bronzite-hypersthene in composition. Grains are equant and anhedral, and are commonly molded on plagioclase and augite; a few are more than 3 mm across, but most are less than 0.5 mm. They have a distinctive colorless to pink pleochroism and patchy extinction. Like olivine, orthopyroxene composition varies consistently through the layer, being most Mg-rich in the center where it is a bronzite with an average composition of En_{75}. Toward the bottom and top of the layer some grains of hypersthene, commonly ranging from En_{70} to En_{65}, have an outer zone containing graphic and lamellar exsolution clinopyroxene. At the base of the layer a few orthopyroxene grains have exsolved cores surrounded by primary hypersthene (En_{61}) without exsolution. Zoning is less extreme in the center of the layer, which is consistent with the small range in the composition shown, and, in fact, many grains appear unzoned. Toward the top of the layer, zoning is more extreme and the range in composition between grains is greater than in the rocks toward the base of the layer.

The monoclinic pyroxene has been shown by spectrographic analysis to be normal augite. Grains are equant and anhedral, and measure from 1 to 2 mm across. A few grains have broad exsolution lamellae.

The opaque iron mineral is mostly ilmenite, which formed anhedral grains that measure from 0.1 to 0.4 mm across. Parts of some grains consist of titanomagnetite. Grains of chalcopyrite, about 0.02 mm across, occur

PHOTOMICROGRAPH OF HYALOSIDERITE DOLERITE, ENGLEWOOD CLIFF

A photomicrograph of a thin section in plane polarized light of the hyalosiderite dolerite (Mg-olivine layer), W-824-60, Englewood Cliff. It shows in the center a plagioclase area of crystallization containing euhedral and subhedral olivine crystals. The grains at the top left and bottom right are orthopyroxene that contains a couple of small serpentine-like pseudomorphs after olivine. The other pyroxene grains at the top and bottom are clinopyroxene which contain a few anhedral olivine inclusions. The olivine grains between the plagioclase and clinopyroxene crystals commonly have euhedral grain boundaries against the plagioclase and anhedral against the pyroxene. In the bottom left hand corner biotite flakes cluster around a few opaque iron mineral grains. The field covers 3 mm x 2.33 mm.

ORTHOPYROXENE IN HYPERSTHENE DOLERITE, ENGLEWOOD CLIFF

An orthopyroxene grain in the hypersthene dolerite, W-U-60, Englewood Cliff Section. The photomicrograph shows in the bottom left center a bronzite core (En_{76}), with patchy extinction, mantled by a broad zone of inverted pigeonite showing typical graphic and lamellar exsolution of the Palisades type. The black fringe (En_{50}) in the top right hand corner is free of exsolution and is in optical continuity with the hypersthene between the exsolved clinopyroxene of the inverted pigeonite zone. The photomicrograph was taken with nicols crossed, and the field covers 2 mm x 1.5 mm.

independently or in association with the Fe-Ti oxides (I. Pontifex, 1965, written commun.).

Strongly pleochroic red-brown biotite occurs throughout, and flakes are more abundant than in the dolerites above and below the layer. Some flakes fringe the opaque minerals.

Bronzite dolerite (3[i]). The dolerite immediately above the Mg-olivine layer is fairly typical of an early stage in the crystallization of a fractionated tholeiitic magma, except that some pigeonite appears to co-exist with primary orthopyroxene. With the abrupt cessation in olivine crystallization there was a corresponding increase in the amounts of orthopyroxene and monoclinic pyroxene crystallizing. The dolerite is subophitic and consists mainly of plagioclase, augite, and bronzite, and contains minor amounts of pigeonite, micropegmatite, opaque iron mineral, biotite, chlorite, and a few grains of a serpentine-like mineral after olivine. Most plagioclase laths are less than 1 mm long; a few range up to 2 mm. Grains are subhedral and have an average composition of An_{61}.

The monoclinic pyroxene formed grains that average about 1 mm across, though a few are up to 3 mm. Much of it, as indicated by spectrochemical determination, is normal augite. The composition of pigeonite has not been determined. Pigeonite formed primary grains on, and in, augite, with which, in some cases, it is in optical continuity. A few grains show fine exsolution lamellae, but most are of primary crystallization and are readily identified by their near uniaxial positive interference figures. During micrometric analysis 3 percent by volume was identified.

Most orthopyroxene grains are composed of a core of bronzite and an outer broad zone containing lamellar and graphic exsolution clinopyroxene. The relict cores have an irregular outline and an average composition of En_{75}; they show typical pink pleochroism and patchy extinction. The outer exsolved zone constitutes between a third and a half of the grain, the rims of which are En_{60}. Similarly exsolved hypersthene mantles some primary augite grains. Most orthopyroxene grains measure up to 1 mm across and a few reach 3 mm.

The opaque iron mineral is evenly distributed throughout, and is probably mostly ilmenite. Most grains measure from 0.1 to 0.2 mm, though a few are up to 1 mm diameter.

Hypersthene dolerite (4). With the advance of fractionation and increasing Fe-enrichment the magma reached a point where crystallizing pigeonite readily inverted (Hess, 1941). Most of the orthopyroxene in the late-stage hypersthene dolerite shows extensive lamellar and graphic exsolution of clinopyroxene. Pigeonite is absent. The rock is still typically subophitic and consists mainly of plagioclase, augite, and hypersthene, together with minor amounts of micropegmatite, opaque iron mineral (largely ilmenite), and biotite. The last two minerals are less abundant than elsewhere in the intrusion.

Most plagioclase laths are in the 0.2 to 1 mm size range, though a few euhedral crystals measure up to 2 mm by 1 mm. Average composition is An_{62}.

Spectrographic determination of the monoclinic pyroxene indicates that it is a normal augite. Grains are mostly less than 1 mm, though a few reach 2 mm. Some are mantled by hypersthene with extensive exsolution. Exsolution lamellae are uncommon in augite and where present are fine and indistinct. However, bladed augite, up to 4 mm in length, contains a typical spine and prominent herringbone structure. Inverted pigeonite formed intimately with augite, in an association corresponding to the relation shown between primary pigeonite and augite in rocks of stage 5.

Orthopyroxene shows extensive exsolution of clinopyroxene, though some of the large grains contain cores of primary bronzite or hypersthene (Pl. 6), whose composition ranges from En_{76} to En_{63}. These grains measure up to 3 mm, whereas most grains are up to 1 mm across.

Pigeonite dolerite (5 [i] and [ii]). With the disappearance of orthopyroxene and inverted pigeonite, the next stage in the fractionated series is typified by primary pigeonite and augite; the two monoclinic pyroxenes are closely associated in their crystallization. The rock is composed mainly of plagioclase, augite, and pigeonite, together with minor amounts of micropegmatite, opaque iron minerals, hornblende, chlorite, biotite, and apatite. Hornblende contains pleochroic haloes surrounding possible zircon grains. The minor constituents, with the addition of sphene, are more abundant in stage 5(ii) rocks, the more Fe-rich pigeonite stage. The ratio of plagioclase to ferromagnesian minerals increases sharply in the pigeonite dolerites (Table 5), though they retain a subophitic relationship to one another. The texture is dominated, however, by areas of plagioclase crystallization consisting of groups of laths which become stumpy by stage 5(ii). The stage 5(i) laths range mainly from 0.3 to 1 mm in length with a few to 2 mm, but in stage (ii) grainsize is fairly even, and equant subhedral grains average about 1 mm. The plagioclase composition, which averages An_{61} in stage (i), becomes progressively more sodic to average An_{56} by stage (ii).

The pyroxene grains are between 0.5 and 2 mm and include some bladed crystals up to 4 mm in length, whereas those in stage (ii) are of more even size, and include groups of grains up to 4 mm across with subindividuals rarely exceeding 2 mm. Spectrochemical determination of augite shows that its composition changed progressively with fractionation: Ca and Mg decreased, and Fe increased markedly compared to the augites in the orthopyroxene-bearing dolerites. Augite is the major pyroxene, and in the early pigeonite dolerites it is slightly brownish and, in part, turbid owing to alteration. Bladed crystals commonly show typical herringbone structure. However, by the late pigeonite stage it has a distinct mauve tint, which also attests to its increased Fe and Ti contents, and indicates, as does pigeonite in this stage, that the monoclinic pyroxenes tend to ferriferous varieties.

The minor pyroxene phase, pigeonite, mostly occurs in augite grains or is molded on them, though in some bladed augite crystals it forms the spine. Its textural relations with augite, and optical continuity between them in places, indicate that these pyroxenes are closely associated in their crystallization. Independent pigeonite grains are fairly uncommon. Pigeonite can be identified by its low positive axial angle, particularly in the early pigeonite stage, and also by its higher birefringence and apparent higher relief (resulting from its more pitted and cracked surface than augite).

Opaque iron mineral is more abundant in stage (ii) rocks; it is probably mostly titaniferous magnetite. Though of variable distribution in the pigeonite dolerites, grains show a large range in size; many are less than 0.5 mm across, but some have a very irregular shape where penetrated by plagioclase laths, and measure up to 4 mm.

Micropegmatite, common to most of the dolerites (except those of the Mg-olivine layer) consists of an interstitial graphic intergrowth of quartz in essentially sodic plagioclase, which is commonly turbid. It occurs exclusively in plagioclase areas of crystallization.

Ferrohypersthene dolerite (6). With fractionation the Fe-rich pigeonite disappears and its place is taken by a pale-green ferroaugite containing, in places, exsolved (?) ferrohypersthene. The rock consists of plagioclase, pale-green ferroaugite, mauve-brown ferroaugite, micropegmatite, and of small amounts of ferrohypersthene, hornblende, and opaque iron mineral. Micropegmatite is more abundant than in rocks of earlier fractionation stages, and so are the minor constituents biotite, apatite, sphene, and zircon. With the change toward more Fe-rich rocks there is a progressive increase in grainsize, though subophitic texture is large retained. The plagioclase laths range up to 4 mm in length, but most, including stumpy grains, average about 2 mm.

The various pyroxenes are grouped mainly into aggregates of grains from 1 to 2 mm across. The two ferroaugites commonly lack distinguishable boundaries between each other. The mauve-brown ferroaugite is a moderately Fe-rich member, probably titaniferous, of the normal augite trend. Alteration is patchy and turbid, and grains are largely free of other mineral inclusions.

Pyroxene relationships are complex, for the pale-green ferroaugite formed both in the pyroxene grain aggregates where pigeonite crystallized in the late pigeonite dolerite, and in the mauve-brown variety along the fine herringbone lamellae. The early formation of this ferroaugite apparently involved the transformation of both the late pigeonite and the mauve-brown ferroaugite, and optical continuity between varieties is not uncommon. The newly-formed grains of pale-green ferroaugite contain numerous small patches and inclusions of hornblende, biotite, and opaque iron mineral, which appear to be residual products of the reaction. They are free of these inclusions where the ferrohypersthene lamellae have exsolved (?). These lamellae (Pl. 7) developed into small anhedral grains which maintain optical continuity with the lamellae. The grains are clear of inclusions and have one good cleavage indicating straight extinction; they show faint yellowish-pink to gray-green pleochroism. Their optical properties (Table 6) indicate that their composition ranges from En_{35} to En_{30}. Rare independent grains are anhedral or subhedral, and they have a coating of hydrous iron oxide at grain boundaries and in cracks. The ferrohypersthene is nowhere in direct contact with the mauve-brown ferroaugite. It is apparently similar in composition and origin to that observed by Poldervaart (1944) in the iron-rich dolerite of the New Amalfi Sheet.

The evenly distributed opaque iron mineral grains are commonly molded on the silicate minerals. They average about 1 mm across, but a few are up to 2 mm. Polished sections show that most anhedral grains are ilmenite and that the subhedral ones are magnetite. But much of the magnetite is transformed, in part, to titanomagnetite and contains regular exsolution of ilmenite along crystallographic directions. Commonly one side of these grains comprises ilmenite in a globular complex resembling a eutectic structure. Chalcopyrite and pyrite are present in roughly equal but minor amounts. Chalcopyrite formed small grains in ilmenite whereas similar sized pyrite grains are at the margins or in the ferromagnesian minerals. Small biotite flakes mainly cluster around the opaque iron mineral grains.

Fayalite granophyre (7[i]). In this stage of fractionation, fayalite granophyre formed, with numerous equant grains of fayalite similar in size to those of the other major minerals. The rock is coarse grained and granophyric. It appears to be missing from the Englewood Cliff and Edgewater Sections, and possibly is faulted out. It is well developed in the Union City Section, where it formed in a horizon about 100 feet thick.[1] Rocks of this Fe-olivine horizon correspond most closely in fractionation to the granophyric dolerite (7[ii]), in the Englewood Cliff Section (Table 8, *compare* analyses of specimens W-WU3-61 and W-F-60).

[1]The type specimen (No. 16441) representing this stage of fractionation came from the Union City Section, as grainsize and alteration precluded the separation of fayalite and apatite from any of the Englewood Cliff rocks.

RELATIONSHIP OF FERROAUGITE TO FERROHYPERSTHENE, ENGLEWOOD CLIFF

A compound ferromagnesian grain in the ferrohypersthene dolerite, W-J-60, Englewood Cliff Section. It shows the incipient formation of ferrohypersthene (light gray) from pale-green ferroaugite (slightly darker gray). Ferroaugite containing fine lamellae (exsolution?) is prominent at the left of the grain. The incipient orthopyroxene forms parallel to these lamellae; it merges, and is in optical continuity with the massive orthopyroxene at the top center and bottom left of the grain. Optical properties of the massive orthopyroxene indicate that its composition is about En_{35}. The photomicrograph was taken with nicols crossed and the field covers .25 mm x .33 mm.

FAYALITE GRANOPHYRE, UNION CITY

A photomicrograph of a thin section in plane polarized light of the fayalite granophyre, W-WU3-61, Union City Section. It shows anhedral fayalite grains to the right of the center and in the bottom left hand corner. Farther to the right laths of apatite can be seen. To the left of the center the large euhedral plagioclase crystal is fringed with micropegmatite showing cauliflower structure. The field covers 3 mm x 2.33 mm.

Stage 7(i) fayalite granophyre contains much more olivine, micropegmatite, hornblende, apatite, and zircon than rocks of stage 6, and obviously has formed in somewhat more hydrous conditions. This gave rise to various alteration products of pyroxene and olivine, which include widespread uralitic amphibole and a rich golden-brown or olive-green serpentine-like mineral (possibly forms of iddingsite or bowlingite). Some of the plagioclase and alkali feldspar in micropegmatitic patches are turbid and sericitized. Much of the ferroaugite seen in earlier stages of fractionation has disappeared, its place being taken by amphibole and biotite. The rock consists mainly of plagioclase, micropegmatite, green and brown hornblende, and fayalite, together with small amounts of opaque iron minerals, green-brown biotite, pale-green and mauve-brown ferroaugite, apatite, and zircon. The pale-green ferroaugite is the more abundant pyroxene.

The grainsize of major minerals is fairly even; grains average about 2 mm across, though some are up to 4 mm, including many of the fayalite grains. Micropegmatite patches are of the same order of size; in many cases, however, micropegmatite mantles isolated euhedral laths and stumpy plagioclase crystals (Pl. 8). Plagioclase shows only slight zoning compared with the rocks of earlier fractionation stages, and has an average composition of An_{37}.

The olivine formed clear pale yellow unzoned anhedral grains with an average composition of Fo_{10}. A visual estimate indicates that about 5 percent by volume of the rock is olivine. Grains are, in general, free of inclusions except for opaque iron mineral in cracks.

The opaque iron mineral grains and biotite flakes are much smaller than the average grainsize of the rock, and some opaque mineral formed skeletal grains. Apatite formed numerous needles and long stout laths.

Granophyric dolerite (7[ii]). This is a variant of the fayalite granophyre (7[i]) and is a dolerite with much less olivine but with a dominantly granophyric texture. It is the most granophyric dolerite encountered in the Englewood Cliff Section. It developed in a fairly hydrous, iron- and alkali-enriched magma horizon. Its texture and grain-shapes indicate that the minerals formed in an environment favoring freedom of crystallization.

The dolerite consists mainly of plagioclase, micropegmatite and quartz, hornblende, and biotite, and contains minor amounts of opaque iron mineral, fayalite, and apatite. A few grains of sphene, pale-green ferroaugite, calcite, and yellow-brown "iddingsite" are also present. Zircon grains are fairly widespread, though small, and are surrounded by pleochroic haloes in biotite and hornblende. The crystals of sphene are strongly pleochroic from light to dark brown. Apatite is much more abundant than in most other dolerites in the fractionation series, and formed typical long euhedral laths. The opaque iron mineral grains are irregular-shaped and skeletal, and like the independent quartz grains are, in general, less than 1 mm across.

Nearly all of the plagioclase is sericitized and turbid, and its average composition is unknown; it formed broad grains up to 4 mm in length. Large areas of micropegmatite crystallization surround many of the plagioclase crystals, including the large ones. Micropegmatite consists of coarse graphic quartz in a turbid base of alkali feldspar, which sodium cobaltinitrite stain test (Chayes, 1952) shows is essentially sodic with only small amounts of potassic feldspar.

The hornblende is the most common dark mineral and occurs as anhedral grains measuring from 0.5 to 2 mm; a few reach 4 mm. Much of it is free of inclusions, and many large grains of green hornblende contain patches of strongly pleochroic brown hornblende in optical continuity, and without distinguishable boundaries between varieties.

Ferrodolerite (8). The ferrodolerites are products of the most Fe-rich stages of fractionation, and are essentially gabbroic, with pyroxene and plagioclase subophitically intergrown in some cases. But mineralogically and chemically, they are distinct from the pegmatitic dolerite of the schlieren, because they formed at a specific stage in the fractionation sequence. The ferrodolerites consist mainly of plagioclase, micropegmatite, quartz, mauve-brown ferroaugite, opaque iron mineral, and pale-green ferroaugite, together with minor amounts of fayalite, sphene, apatite, and zircon. Hydrous conditions during crystallization led to the development of much green-brown hornblende, biotite, and yellow "iddingsite;" golden-brown and green biotite commonly cluster in small flakes near opaque iron mineral.

The plagioclase formed fresh subhedral grains that measure between 1 and 5 mm across. It is zoned and complexly twinned, and has an average composition of An_{47}.

Ferroaugite is almost entirely the mauve-brown variety, and the depth of color suggests that it is probably titaniferous. It formed equant grains up to 5 mm across and bladed crystals to 10 mm in length. Like plagioclase, few grains are less than 1 mm, and most are in the 2 to 3 mm range. Fine herringbone lamellae are common, and they are particularly marked in bladed crystals with a spine. The pale-green ferroaugite appears to have developed initially along these lamellae in the central zone, as well as at grain margins, and as it did, it maintained optical continuity with the original ferroaugite grain. With the change, the pyroxene became crowded with minute inclusions, mainly hornblende, bulsome are finely divided iron oxide. With further reaction, the inclusions were progressively resorbed as the green-brown hornblende developed.

Micropegmatitic residuum comprises a major proportion of the rocks. Much of it is interstitial to plagioclase crystals, and the quartz inclusions, which are in an essentially turbid alkali feldspar base, developed cauliflower-shaped and graphic patterns. The base also contains small sericite, chlorite, and biotite flakes; apatite normally occurs in the micropegmatite areas as long needles.

Small quantities of fairly granular fayalite (Fo_7) fringe some opaque iron mineral grains. Much of it, however, has been altered to a variety of colored serpentine-like minerals, lemon-yellow, golden-brown, and green, all of which are poorly anisotropic, and are probably forms of iddingsite or bowlingite.

Opaque iron mineral grains are abundant, and have very irregular margins or form skeletal structures. Half of the opaque mineral is titanomagnetite, and most of the remainder is ilmenite. Small chalcopyrite grains are disseminated through the rock and some are enclosed by the ilmenite. Blebs of Fe-Ti oxides and chalcopyrite formed graphic-like intergrowths with pyroxene.

Pegmatitic dolerite. As the intrusion crystallized, parts of the volatile-enriched liquid became isolated from the main fractionating magma, and formed pod-like masses or schlieren of pegmatitic dolerite. Schlieren are randomly distributed, but are most common in the top third of the intrusion.

The dolerite is typically pegmatitic and subophitic, and is characterized by bladed mauve-brown ferroaugite crystals, many of which exceed 10 mm in length; these show herringbone structure, and many have spines or inclusions of pigeonite. In field exposures it is not uncommon to see bladed pyroxene crystals up to 5 cm in length. Between the large pyroxene crystals are broad laths and equant grains of plagioclase up to 4 mm across, and some micropegmatite and quartz. Hornblende is also an important constituent. Opaque iron mineral, sphene, apatite, and zircon are accessory. Apatite needles and opaque iron mineral grains, range up to 4 mm in length. Nowhere has fayalite been seen, though a clear pale-green pyroxene has formed along the herringbone lamellae of the mauve-brown ferroaugite in places. With it, myriads of small mineral inclusions developed in the pyroxene. Both pyroxene and plagioclase show patchy turbid alteration, which, together with uralite and the part-transformation of pyroxene to green and green-brown hornblende, indicate crystallization was under fairly hydrous conditions. Grain boundaries of opaque iron mineral are very irregular, and commonly are fringed by sphene.

Pigeonite dolerite. This dolerite is very similar to that described from stage 5(i) of the type section, but, with the bronzite dolerite described next, provides critical evidence for the internal chilled contact at Haverstraw Quarry. These two rocks indicate a reversal in pyroxene crystallization which resulted from an abrupt change in physical conditions during the consolidation of the intrusion.

The pigeonite dolerite is subophitic and consists of plagioclase, augite, and pigeonite, and of accessory opaque iron mineral, micropegmatite, and turbid mesostasis. Both augite and plagioclase show patchy turbid alteration. Minor sericitization and uralitization also occurred. A few small patches of serpentine-like mineral fringed by opaque iron mineral in augite grains are reminiscent of olivine pseudomorphs. Opaque iron mineral grains range up to 1 mm across and include a few skeletal grains.

The plagioclase is typically zoned and twinned, and has an average composition of An_{63}. Laths measure up to 2.5 mm, but most are between 0.5 and 1 mm in length.

About two-thirds of the pyroxene comprises an augite that ranges from 0.4 to 1 mm across, but includes bladed crystals up to 3 mm by 0.5 mm. The remaining third is pigeonite, prisms of which measure from 0.3 to 1.5 mm. The pigeonite, though intimately associated with augite in many cases, can be clearly distinguished by its apparent higher relief, pitted and irregularly cracked surface, and common inclusions of opaque iron mineral (Pl. 9). Appropriate sections give near-uniaxial positive interference figures. Compositions of the pyroxenes have not been determined.

Bronzite dolerite. The mineralogical composition of this bronzite dolerite sharply contrasts that of the pigeonite dolerite (*compare,* Pls. 9 and 10). The bronzite dolerite is fairly typical of an early dolerite fractionation stage. It is subophitic and contains large bronzite grains that measure up to 4.5 mm by 1.5 mm; these are possibly slightly resorbed microphenocrysts. The dolerite consists of plagioclase, augite, and bronzite, but contains minor amounts of pigeonite and accessory opaque iron mineral and turbid mesostasis with rods of opaque mineral. Mesostasis is more common than in the pigeonite dolerite, and some patches comprise quartz and alkali feldspar in a fine cauliflower-like structure. Mineral grains do not show flow orientation, but many attempts to measure bronzite grains by U-stage were abortive because β was along the axis of the diamond drill core from which thin sections were cut.

The plagioclase laths range from 0.2 to 1 mm, but a few crystals are stumpy and measure up to 2 mm across. Some of the large crystals have partly sericitized cores mantled by plagioclase showing reverse zoning, and in turn, by normal zoned plagioclase. The average composition is An_{66}; the most basic crystals form small inclusions in pyroxene.

Most of the monoclinic pyroxene is normal augite, grains of which range up to 0.5 mm, and a few bladed crystals reach 1.5 mm in length. The augite is zoned in some cases and early pyroxene formed cores including pigeonite. The pigeonite, though much less abundant than augite, is a primary variety

PIGEONITE DOLERITE, INTERNAL
CHILLED CONTACT, HAVERSTRAW

A photomicrograph of a thin section in plane polarized light of the pigeonite dolerite, 14-105, Haverstraw Quarry. It shows in the center a pair of pigeonite grains and beneath them an area of turbid glassy mesostasis. Other minerals are augite, plagioclase, and opaque iron minerals. The field covers 3 mm x 2.33 mm.

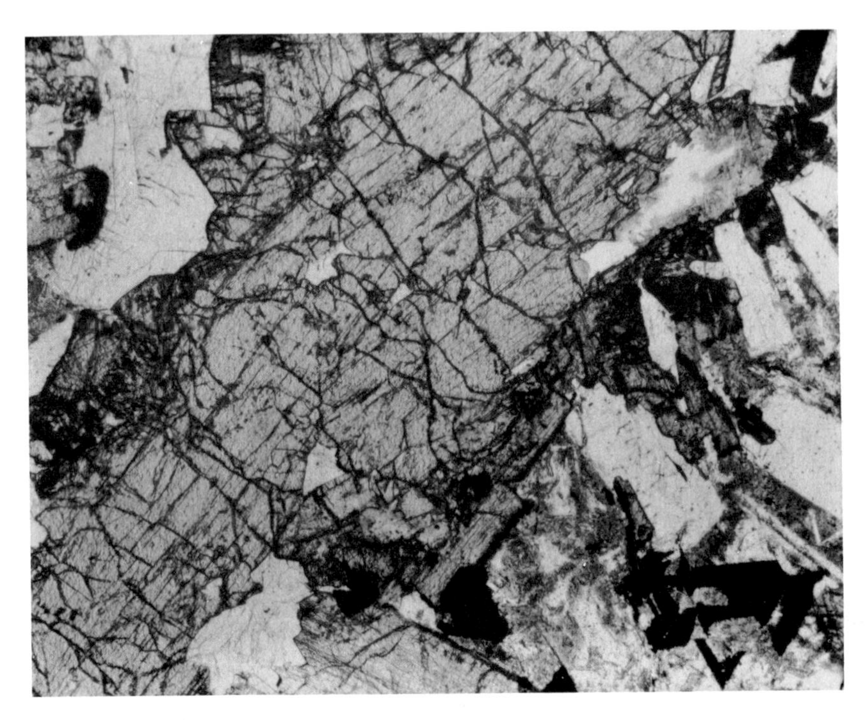

BRONZITE DOLERITE, INTERNAL
CHILLED CONTACT, HAVERSTRAW

A photomicrograph of a thin section in plane polarized light of the bronzite dolerite, 14-109½, Haverstraw Quarry. It shows a large bronzite grain ($En_{77.5}$) with augite and a few pigeonite grains molded on it. The turbid area between the opaque iron mineral grains in the lower right hand quadrant is glassy mesostasis. The laths are plagioclase. The field covers 3 mm x 2.33 mm.

intimately associated with augite crystals, and has similar distinguishing features to that in the pigeonite dolerite. A few grains are slightly exsolved.

Most orthopyroxene grains are large and have very irregular grain margins. Augite and a few pigeonite grains envelop many of them (Pl. 10). The bronzite has an average composition of $En_{77.5}$; it is unzoned and shows typical patchy extinction and pink pleochroism. A few inclusions of serpentine-like mineral with opaque iron mineral appear to be pseudomorphs of corroded olivine grains.

Late chilled dolerite. The Palisades Sill was intruded by dikes after it had solidified. An example of these can be seen in the Union City Section, where a 60-foot wide dike of chilled bronzite dolerite and another about 4 feet wide intrude the sill near Central Avenue, apparently exploiting a fault that brought dolerite of stage 5(ii) against that of stage 6.

The bronzite dolerite is integranular and has a few microphenocrysts of plagioclase, augite, and orthopyroxene. The smaller of the two dikes also contains glomeroporphyritic aggregates of augite and pigeonite up to 3 mm across, containing subindividuals from 0.2 to 0.6 mm. These may, however, be xenocrysts incorporated from the wall rocks. Olivine is absent from both dikes. Opaque iron mineral and biotite are accessory. The groundmass consists of anhedral augite and orthopyroxene (augite being in slightly greater abundance), and of plagioclase laths. The grainsize of the base averages about 0.2 mm.

The plagioclase laths are fresh and range up to 0.5 mm, but most are between 0.1 and 0.2 mm in length. A few grains are stumpy.

The clinopyroxene is augite; most grains are equant and anhedral, and measure between 0.05 and 0.3 mm. The crystal form of some large grains is outlined by zones of opaque mineral inclusions.

The orthopyroxene is present as euhedral microphenocrysts up to 1 mm in length, and as subhedral and anhedral grains averaging about 0.2 mm. It shows distinct pink pleochroism, and is without zoning, patchy extinction, or exsolution. Some of the large grains are fringed by augite that contains abundant opaque inclusions.

The opaque iron mineral is evenly distributed throughout. A few grains reach 0.4 mm across, though most are much smaller and generally less than 0.05 mm, particularly the abundant inclusions in some augite grains. In addition, opaque iron mineral grew along the crystallographic planes of some augite grains and formed delicate lattice structures. Small brown biotite flakes are commonly associated with the opaque mineral, but are also interstitial to the other minerals.

Late sodic veins. Narrow and sinuous residual Na-rich veins with aplitic texture formed at various horizons in the intrusion. They are of late hydrothermal origin, and consist mainly of turbid oligoclase and quartz, but also contain analcite and a few grains of diopside, clinozoisite, sphene, calcite, sericite, chlorite, and opaque iron mineral.

SCHEMATIC DIAGRAM OF MINERAL SERIES

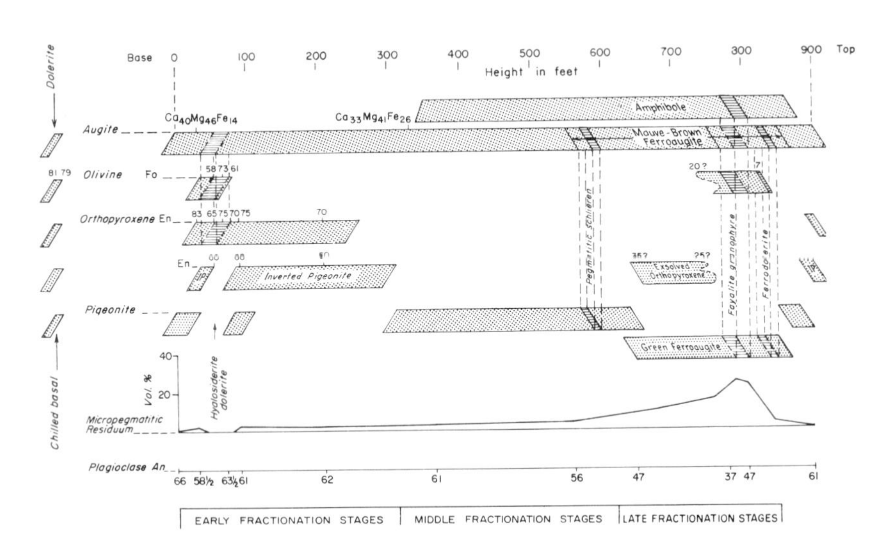

Figure 6. Schematic diagram showing the distribution and variation in the composition of the main mineral series in the Palisades Sill. It should be viewed from the center and read from left to right. The diagram has been compiled to represent the Englewood Cliff Section, and the distances over which the various minerals crystallized have been measured in this section. However, because of the incompleteness of any one section, it is necessarily, in part, a composite diagram. The information on olivene and orthopyroxene in the Mg-olivine layer has been obtained from the Edgewater Section, and that on the fayalite granophyre has been interpreted from the Union City Section. The location and size shown for the pegmatitic schlieren were chosen arbitrarily.

MINERALOGY

The main primary minerals in the Palisades dolerite are plagioclase, monoclinic pyroxenes, orthopyroxenes, and olivines. Micropegmatite, hornblende, and opaque iron minerals are important constituents in the top third of the intrusion. The Fe-Ti oxides occur throughout, however, and with biotite and quartz are the major accessory minerals; others are apatite, zircon, chalcopyrite, and sphene. Movement of volatiles caused local alteration to minerals at all stages in fractionation. Patchy sericitization and turbid alteration of feldspars, uralitization of pyroxenes, and alteration of olivines to serpentine-like minerals (including varieties of iddingsite and bowlingite) and to iron oxides, are not uncommon.

Only the main mineral series will be dealt with, except for brief comments on the accessory minerals and alteration products. The discussion of variations in mineral composition with fractionation is limited to major elements and to the main cations that enter into mineral lattices in small amounts. The distribution of trace elements in mineral phases will be treated in a separate section. The schematic diagram, Figure 6, summarizes the mineralogy of the main mineral series and shows the distribution, and variation in composition, of each mineral from bottom to top in the intrusion. The optical properties and composition of minerals important to petrological and chemical discussion are presented in Table 6. Compositions are reported in molecular percent.

The importance of a complete understanding of the monoclinic pyroxenes to the study is fully recognized. Many attempts were made to isolate pigeonite from augite for chemical work, but without success. Study of these important series of minerals will require more time, and the results reported here are of a general investigation only. More time could also be given with advantage to the study of the opaque iron minerals.

Pyroxene

The principles operating in pyroxene crystallization in the Palisades Sill were similar to those in other fractionated oversaturated tholeiitic intrusions, but the pyroxene fractionation trends were complicated, it is believed, by an

Table 6. Mineralogical Determinations

			PLAGIOCLASE			OLIVINE	
Specimen	Section	Height*	R.I.	An[1]	$2V_X$[2]	Fo[4] (mol %)	Comment
W-889LC-60	Englewood	1	1.5425	66		Range 81.5-79[5]	Unzoned Microphenocrysts 20 determinations
W-865-60	Englewood	30	1.5355	58.5			
W-OLO-61	Edgewater	55				Range 63-55[5] Av. 58 Mode 57.5	Zoned 90 determinations
W-OL127-61	Edgewater	66				Range 77-69[5] Av. 73 Mode 73.5	Zoned 50 determinations
W-824-60	Englewood	70	1.540	63.5		73[6]	
W-OL275-61	Edgewater	78				Range 69-58[5] Ave. 61 Mode 58.5	Zoned 16 determinations
W-804-60	Englewood	90	1.5375	61			
W-U-60	Englewood	215	1.539	62			
W-R-60	Englewood	365	1.538	61			
W-N-60	Englewood	560	1.533	56			
W-J-60	Englewood	685					
W-E-60	Englewood	805	1.525	47			
16441	Union City	≈980	1.517	37	54°, 54, 53.5		Unzoned
W-WU3-61	Union City	990			54°, 54, 54, 53.5, 53.5, 52, 52	Range 11-7 Av. 10	Unzoned
14-104	Haverstraw	102	1.5395	63			
14-109½	Haverstraw	108					
14-112	Haverstraw	110	1.5435	66			

(1) Average composition obtained from R.I. of fused plagioclase. Schairer and others (1956, Fig. 28)
(2) Universal Stage—double optic axis determination
(3) Universal Stage—single optic axis determination

TABLE 6. (CONTINUED)

ORTHOPYROXENE				AUGITE	
2Vx[3]	2Vx[2]	En[4] (mol %)	Comment	Composition [6]	Comment
85°, 79, 79, 79		Range 85-82 Av. 83	Unzoned Patchy extinction No exsolution	$Ca_{40}Mg_{46}Fe_{14}$	Approximate composition only
62°, 62, 58, 58, 58, 56, 53 53°	59°, 58, 54, 53, 53	Range 70-60 Av. 65 Rim 60	Zoned Some exsolved grain centers and patchy extinction		
70°, 68, 66, 66	68°, 67, 67, 67	Range 76-73 Av. 74½			
66°, 58, 58, 58	64°	Range 73-67 Av. 69 70[6]	Some crystals show slight zoning and patchy extinction	$Ca_{39}Mg_{45}Fe_{16}$	"
77°, 75, 66, 56	72°, 60, 58, 58, 58, 55, 54	Range 81-62 Av. 70	Some crystals show zoning and patchy extinction Exsolution rare		
76°, 71, 71, 62, 60	71°, 69, 69, 66 53°, 53	Range 80-68 Av. 75 Rim 60	Some crystals zoned and show patchy extinction Exsolved grain margins common	$Ca_{38}Mg_{45.5}Fe_{16.5}$	"
68°, 59, 55	69°, 68, 60, 58 48°, 48, 48	Range 76-63 Av. 70 Rim 50 58[6]	Zoned Exsolution abundant	$Ca_{38}Mg_{44}Fe_{18}$	"
				$Ca_{33}Mg_{41}Fe_{26}$	"
				$Ca_{34}Mg_{36.5}Fe_{29.5}$	"
	62°, 61, 60, 59, 58, 57	Range 35-30 Av. 32.5	Unzoned		
				$Ca_{31}Mg_{33}Fe_{36}$	"
73°	68°	Range 79-75 Av. 77½	Unzoned No exsolution		
	75°, 72				

(4) Deer and others (1962, v. 1, Fig. 11; 1963, v. 2, Fig. 10)
(5) Electron probe determination (Anal D. H. Green)
(6) Optical spectrograph determination
* Height above base in feet

injection of a second phase of magma during the crystallization of the first. This modified the physico-chemical conditions within the intrusion, and caused some overlap and reversals in trends normally expected in a simple intrusion.

During much of the fractionation of simple tholeiitic intrusions two pyroxenes crystallize in cotectic equilibrium (Ca-rich and Ca-poor varieties) and in the latest fractionation stages, beyond the two-pyroxene field, usually a single pyroxene phase only has been accepted, though some workers have reported that two variants of the ferroaugite phase occur. The trend of pyroxene crystallization in the Palisades Sill agrees with these findings, except that, beyond the two-pyroxene field, there appear to be two distinct ferroaugites and a Fe-rich orthopyroxene. The compositional trend of all the pyroxenes during fractionation is to successive enrichment in Fe^{2+} relative to Mg, and to some variation in Ca content.

Augite. The Ca-rich pyroxene is a normal colorless augite, which, with fractionation, gradually changed to a mauve-brown ferroaugite, probably a titaniferous variety. Typical twinning occurs in places. In the Mg-olivine layer a few augite grains contain broad exsolution lamellae, but in the other dolerites lamellae are fine and indistinct, except in the pigeonite dolerites and ferrodolerites, where typical herringbone structure is common. Alteration is patchy and turbid. The Ca-rich monoclinic pyroxene crystallized in all fractionation stages except for a few rocks of late hydrothermal origin, where its place is taken by amphibole.

Augite distribution in the intrusion is shown in Figure 6. Also shown are the two lines of pyroxene evolution with fractionation. Normal augite fractionates continuously to a mauve-brown ferroaugite, whereas the pale-green ferroaugite appears to evolve in late fractionation stages from the orthopyroxene-pigeonite trend. This ferroaugite occurs in the pyroxene grain aggregates in a manner similar to pigeonite of the late-middle fractionation stages. But pyroxene relationships are complex, as the formation of the pale-green ferroaugite appears also to involve the mauve-brown ferroaugite in the reaction and both varieties co-exist throughout the late fractionation stages.

The chemical investigation of the pyroxenes was by optical emission spectrograph, and the calculated compositions from the determinations of major elements can only be regarded as approximate. The first augite to crystallize, which co-existed with orthopyroxene, has a composition of roughly Ca_{40} Mg_{46} Fe_{14}[2]. In the late pigeonite stage of fractionation the augite has a composition of approximately Ca_{34} Mg_{36} Fe_{30}, and in one of the Fe-rich stages the composition of the mauve-brown ferroaugite is about

[2]Compositions refer to molecular percent and Fe = Fe^{2+} + Fe^{3+} + Mn.

Ca_{31} Mg_{33} Fe_{36}. There is an appreciable increase in Fe and a corresponding decrease in Ca and Mg, though the variation does not appear to have been as great as it is in some other measured augite series, as, for example, in the Skaergaard (Muir, 1951; Brown, 1957). However, the composition of the pale-green ferroaugite has not been determined.

Trace element determinations by the optical emission spectrograph are more accurate than the major element determinations, and values obtained for Cr, Ni, and Sr in the clinopyroxene of the Mg-olivine layer (Table 14) indicate that it contains less Fe, more Mg, and somewhat less Ca than the clinopyroxenes immediately above and below the layer. From these trace element results, together with the major element values obtained by optical emission, the clinopyroxene compositions may be tentatively re-interpreted as follows:

W-865-60	Ca_{40} Mg_{45} Fe_{15}
W-824-60	$Ca_{39\cdot5}$ Mg_{46} $Fe_{14\cdot5}$
W-804-60	Ca_{40} $Mg_{44\cdot5}$ $Fe_{15\cdot5}$.

Detailed work on the clinopyroxenes will probably show that their compositions vary sympathetically with those of the orthopyroxenes and olivines in the layer, and that clinopyroxene is more Fe-rich in the late fractionation stages than the present study indicates.

The Al content of the augites in the Palisades Sill is fairly comparable to that of analyzed clinopyroxenes of other intrusions (*see* Hess, 1949; Muir, 1951; Kuno, 1955; Brown, 1957). It shows a slight decrease with fractionation, which is also consistent with the observations of Brown (1957) on the Skaergaard.

With fractionation, the amount of Ti in the augites increases; this variation agrees with the findings of Walker and Poldervaart (1949) on the Karroo, of Hess (1960) on the Stillwater, and of McDougall (1961) on Red Hill. In the Skaergaard, Brown (1957) found little change in Ti content with fractionation, and Ti values are greater than those for the Palisades Sill and Red Hill.

The amount of Mn in the augites of the Palisades Sill is comparable to that in the Skaergaard and Stillwater intrusions, though Mn shows a steady increase with fractionation following Fe^{2+}, as it does in Red Hill and the other intrusions.

Orthopyroxene. A few microphenocrysts of orthopyroxene are in the lower and upper chilled dolerites. Orthopyroxene crystallized mainly, however, in the lowest quarter of the intrusion, and where the fractionation sequence reverses toward the top contact. In addition, small amounts of a Fe-rich variety formed in the late fractionation stages.

Two varieties of orthopyroxene are present in the lower part of the intrusion. The orthopyroxene that crystallized directly from the magma has been called "Bushveld type" by Hess (1960, p. 23). That derived by inversion of pigeonite is a variant of the "Stillwater type" (1960, p. 27 and p. 34) which Hess has called "Palisades type." This type characteristically developed in hypabyssal intrusions of dolerite where cooling was sufficiently rapid to prevent the complete diffusion of Ca^{2+} and the formation of well-defined lamellae of diopsidic clinopyroxene before inversion. As a result the excess Ca^{2+} was not arranged regularly through the hypersthene structure, and minute blebs of the clinopyroxene were expelled all through the orthopyroxene. In most cases poorly defined fine lamellae and blebs in a rough cuneiform structure formed, which may, or may not, show a tendency toward elongation on the (001) or parallel to the (100) plane. The Palisades-type orthopyroxene is illustrated in Plate 6.

The primary orthopyroxene of Bushveld type without exsolution of clinopyroxene shows clear pale-green to pink pleochroism and typical patchy extinction (Pl. 6). The Palisades-type orthopyroxene originally crystallized as pigeonite, and later inverted to the orthorhombic form. Pleochroism is less marked and patchy extinction is absent. The Fe-rich variety that appears to exsolve from the pale-green ferroaugite forms small anhedral and subhedral grains of late crystallization. These commonly have good cleavage and slight pleochroism from gray-green to yellowish-pink tints. With fractionation the various types became progressively richer in Fe. The range of crystallization of the orthopyroxenes and variations in their compositions are shown in Figure 6, and their optical properties are given in Table 6.

Some of the orthopyroxene in the basal chilled dolerite appears to have crystallized from the magma after emplacement. It formed grains enveloping other minerals and contains exsolved clinopyroxene. Similar orthopyroxene formed as reaction rims to some olivine microphenocrysts. However, a few microphenocrysts of Bushveld-type orthopyroxene occur in the glomeroporphyritic aggregates of pyroxene in the chilled contact dolerite, and in dolerite of fractionation stage 1 where its composition is En_{83}. Similar orthopyroxene is preserved as relict cores in some of the augite microphenocrysts. These orthopyroxene grains are unzoned, and in the stage 1 dolerites, augite and a few pigeonite grains commonly have built on their somewhat corroded grain margins.

At the base of the Mg-olivine layer the composition of the Bushveld-type orthopyroxene ranges from En_{70} to En_{60} between grains, and at the top from En_{81} to En_{62}; in each case many grains are zoned. In the center of the layer the composition range is En_{76} to En_{67}, and zoning is only slight and exsolution absent. The pattern of exsolution also changes from bottom to top of the layer. At the base exsolution of clinopyroxene occurs only in primary orthopyroxene grains; it is patchy, and patches commonly form the cores of grains, though some occur to one side of the grain or in a central band. In a few cases, primary orthopyroxene grains have exsolved cores surrounded by a broad zone of hypersthene (En_{61}) without exsolution structure. As mentioned, exsolution is rare in orthopyroxene in the center of the layer; but at the top, exsolution of clinopyroxene progressively increases, and by stage 3 (i) most orthopyroxene grains have bronzite cores (Av. En_{75}) surrounded by a broad zone of Palisades-type orthopyroxene (En_{60}) which constitutes from .33 to .50 of the grain. The variation in composition of Bushveld-type orthopyroxene from the bottom to the top of the layer is shown in Figure 7.

In fractionation stage 4, the stage before that in which pigeonite remained the stable phase, most of the orthopyroxene was derived by inversion of pigeonite, and grains contain much exsolved clinopyroxene. A few of these grains, however, have Bushveld-type orthopyroxene cores whose composition ranges from En_{76} to En_{63} (Pl. 6).

Hess (1941, p. 581) observed that the inversion of pigeonite to hypersthene took place when the molecular ratio Fe/Mg of 3:7 was reached in mineral composition. With fractionation in the Palisades Sill the orthopyroxene shows an over-all increase in Fe/Mg ratio, but, as mentioned, the course of the trend has a reversal in the Mg-olivine layer, which indicates that the history of crystallization is a complicated one. Though the composition of some orthopyroxene of primary crystallization is more Fe-rich than En_{70}, and extreme compositional ranges are shown by some orthopyroxene grains as illustrated in Plate 6. En_{70} can be considered the average composition at which inverted pigeonite began to appear in the fractionation sequence.

The present investigation indicates that the composition of primary orthopyroxene ranges from En_{83} to En_{60}, whereas that of the Palisades type ranges between En_{50}, but is mostly between En_{60} and En_{50}. Thus En_{60} was the composition at which othropyroxene of the Palisades type most commonly formed, and En_{50} was the composition reached before pigeonite remained stable through cooling.

According to Hess (1960, p. 29), the bulk chemical composition of the Palisades-type orthopyroxene is similar to that of pigeonite; its Ca content is about 3 times that in ordinary Bushveld-type orthopyroxene. Calcium amounts to about 9.5 percent of the total Ca + Mg + Fe. The content of minor elements and their variation with increasing Fe content is consistent with a continuation of values found in orthopyroxene of the Bushveld type.

The Al content of the orthopyroxene is somewhat less than that of the co-existing clinopyroxene, and about twice that of the olivine (Table 14). It is consistent with other analyzed orthopyroxenes, (Hess, 1960, Tables 3 and 4, p. 25 and 28), and it remains virtually constant with fractionation at about 1.6 percent Al_2O_3. Titanium concentration decreases slightly with fractionation, whereas Mn increases slightly. Manganese probably follows Fe^{2+}.

Spectrographic analyses for Ca indicate that the Bushveld-type orthopyroxene contains 2.5 percent CaO and that the Palisades type has 4.0 percent CaO. Comparing these values with published analyses (Hess, 1960) shows that the Palisades type is similar but that the Ca value obtained for the Bushveld type may be a little high. The Ba value for the Bushveld type is less than 35 ppm (the detection limit of the analytical method used), which, when compared with the 70 ppm in the Palisades type, suggests that CaO in the Bushveld type does not exceed 2 percent, because Ba follows Ca in its distribution in the orthopyroxenes.

After the disappearance of the Mg-rich varieties with fractionation, orthopyroxene did not reappear until the Fe-rich fractionation stages, where it exsolved(?) initially from the pale-green ferroaugite. Optical measurements on the ferrohypersthene dolerite indicate that its composition is between En_{35} and En_{30}.

Pigeonite. Pigeonite forms a discontinuous mineral series in the intrusion. At the Mg-rich end of the fractionation series it has an inversion relationship to orthopyroxene of the Palisades type, and at the Fe-rich end, it appears to give way to the pale-green ferroaugite. The distribution of pigeonite in the intrusion is shown in Figure 6. A detailed chemical and optical study of this important mineral series has yet to be made; it is reasonable to assume, however, that pigeonite shows progressive enrichment in Fe^{2+} and a corresponding decrease in Mg with fractionation.

In most pigeonite-bearing dolerites, the pigeonite formed in intimate association with the Ca-rich monoclinic pyroxene, either as spines or cores to augite grains, or molded on augite. Much less commonly it formed independent grains, and where it co-exists with primary orthopyroxene, it can also, with augite, be molded on grains of orthopyroxene (Pl. 10).

Pigeonite formed as a metastable phase in the chilled-contact dolerite. Beyond the contact, however, it continued to crystallize in small amounts, co-existing with orthopyroxene, and in some cases olivine, nearly to the base of the Mg-olivine layer. It did not crystallize in the Mg-olivine layer, but it did reappear immediately above, where it crystallized over a short distance, again as a primary phase co-existing with orthopyroxene. Thereafter, only the inverted form occurred until a composition range had been reached in the course of fractionation where pigeonite remained stable through cooling.

Continuous pigeonite crystallization followed in stage 5, the middle fractionation stage, which has been subdivided on the basis that dolerite in 5(i) contains pigeonites whose properties are typical of Mg-rich varieties, whereas that in 5(ii) has typically Fe-rich pigeonites. In the next fractionation stage the place of pigeonite appears to be taken by the pale-green ferroaugite.

Elsewhere, pigeonite occurs only in rocks where the fractionated series reversed toward the top contact and in occurrences of pegmatitic schlieren, where it crystallized in association with the mauve-brown ferroaugite.

Olivine

Bowen and Schairer (1935) in their investigation of the system MgO-FeO-SiO_2 postulated a gap in the crystallization of the olivine series. They also showed that with strong fractionation, even in a silica oversaturated melt, an Fe-rich olivine may form in equilibrium with quartz and Fe-rich pyroxene.

Crystallization of olivine in the Palisades Sill agrees in general with these findings, but olivine occurs in three distinct horizons in the intrusion: in dolerite of the basal and top chilled contacts where it indicates that olivine was apparently crystallizing in the magma before emplacement; in the Mg-olivine layer, where it is believed to be a concentration of the initial phase to crystallize after the emplacement of a second magma phase into the partly crystallized first phase; and in the Fe-olivine horizon where it appears to have formed in late stages as a result of normal fractionation in the intrusion. Crystallization of the olivine series and its variation in composition with fractionation are shown in Figure 6; details of composition and optical mineralogy are given in Table 6.

The olivine in the chilled-contact dolerite formed only microphenocrysts, some of which are corroded and may be partly or completely altered to serpentine or have an orthopyroxene reaction rim. Its composititon ranges from $Fo_{81.5}$ to Fo_{79} at Englewood Cliff; individual grains are unzoned.

The Mg-olivine layer ranges in thickness from 2 to 30 feet. Scattered corroded grains appear, however, as much as 20 feet below. The base of the layer is delineated by a rapid increase of olivine to about 10 percent. The top of the layer is sharp; apparently olivine crystallization ceased abruptly, and scattered grains are limited to within a couple of feet above the layer.

The small olivine grains (which by volume are about three times as abundant as the large ones) are generally euhedral or subhedral, and occur mainly in plagioclase areas of crystallization, which give the rock a poikilitic texture (Pl. 5). Individual grains are clear and show normal zoning. The large grains, however, are inclined to be corroded and have irregular cracks with some iron oxide and serpentine-like alteration; most have anhedral boundaries with adjacent ferromagnesian minerals and subhedral with plagioclase (Pl. 5). Normal zoning is also common in the large grains, but zoning in all grains, as for composition range between grains, is less extreme in the center (Fo_{77} to Fo_{69}) of the layer, where some grains are unzoned, than it is in the base (Fo_{63} to Fo_{55}). The greatest range in composition is shown by grains at the top (Fo_{69} to Fo_{58}) of the layer. The variation in composition of olivine from bottom to top of the layer can be seen in Figure 7. Compositionally there is nothing to distinguish the large grains from the small; the composition ranges overlap, though over-all the large grains appear to contain somewhat more Fe than the small.

After its disappearance above the layer, olivine did not reappear in the course of fractionation until the Fe-rich orthopyroxene stage of crystallization was completed. With fractionation, pigeonite presumably gave way to a pale-green ferroaugite and some of this appears to exsolve ferrohypersthene. Possibly in the very late stages this orthopyroxene reacted with liquid to form fayalite and quartz. Fayalite crystallized over a vertical thickness of about 100 feet in the intrusion. It is a clear yellow-tinted variety which may exhibit two cleavages at right angles and irregular fractures. Grains are anhedral, and in any one rock are unzoned. With fractionation, however, the composition progressively changed from $Fo_{20?}$ to Fo_{7}. It is most abundant in the 100-foot zone of the fayalite granophyre (Pl. 8). In the hydrothermal stage, olivine may be altered to green bowlingite or to yellow-brown "iddingsite."

There is a gap, therefore, in the olivine crystallization sequence from Fo_{55} to $Fo_{20?}$. This is consistent with the experimental findings referred to above, and with observations made on other intrusions. Poldervaart (1944, p. 94) observed a break between Fo_{50} and Fo_{20} in the New Amalfi Sheet. Furthermore, in the Palisades, pigeonite does not co-exist with olivine, except in the dolerite below the Mg-olivine layer where scattered olivine grains occur; but these may have settled from overlying magma, as some micropegmatite also occurs in the early dolerite below the layer.

Aluminum in olivine is only about half that in the orthopyroxene, and it shows a very slight increase with fractionation (Table 14). The titanium content in olivine shows a strong increase, from a moderate trace in the hyalosiderite to nearly 0.78 percent TiO_2 in the fayalite, and Mn nearly doubles over the same range to 0.72 percent MnO.

Plagioclase

Plagioclase occurs throughout the intrusion. In the chilled contact dolerite it is a major groundmass mineral, and, in addition, at some localities it occurs in glomeroporphyritic aggregates. In the subophitic dolerites it formed laths and stumpy grains, whereas in the gabbroic Mg-olivine layer the large subhedral crystals are commonly poikilitic with abundant small olivine inclusions. Throughout the intrusion, plagioclase is strongly zoned and extensively twinned; twinning on all common laws has been observed. Zoning is normal and becomes less extreme with fractionation; in a few cases there is reverse zoning, as in the bronzite dolerite of the internal chilled contact at Haverstraw. Alteration is slight and patchy; localized hydrous conditions led to some extensive sericitization and turbid alteration in places.

Walker (1940) has given much detail on plagioclase grainsize measurements and on micrometric variations of plagioclase across the intrusion. This has been discussed further by Walker (1957) and Joplin (1957, p. 131). For the present investigation the variations in composition were established by a

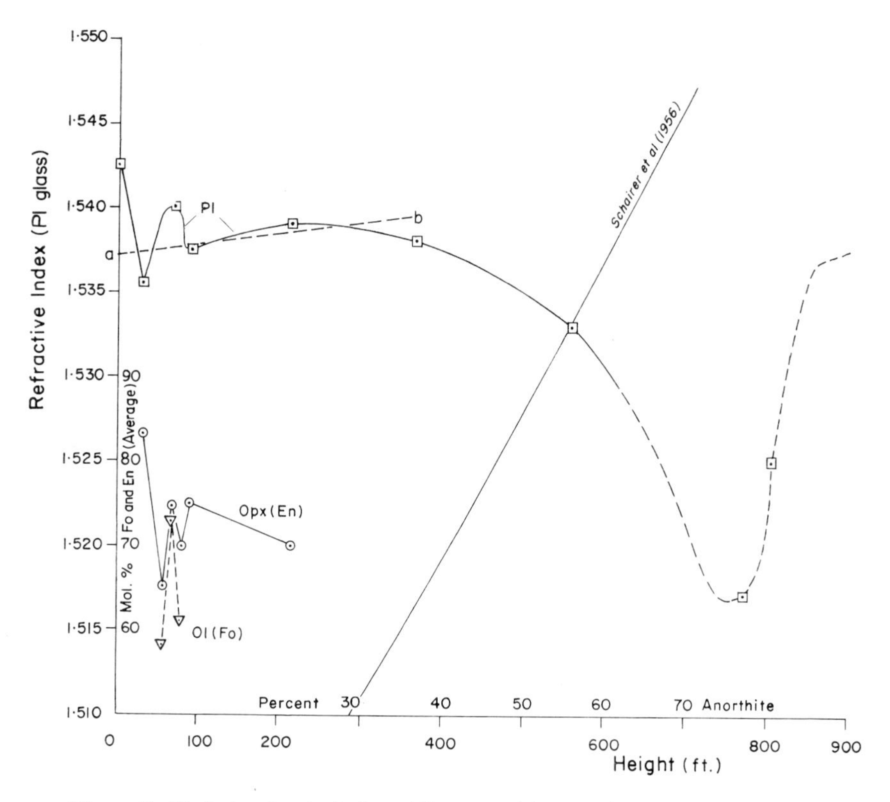

Figure 7. Variation in plagioclase (Pl) composition, Englewood Cliff Section, and in olivine (Ol) and orthopyroxene (Opx) compositions in early fractionation stages, Edgewater Section.

study of the average plagioclase composition of the type specimens that represent the various stages of fractionation. The results differ significantly from those given by Walker (1940, p. 1070, Fig. 4), which were obtained by conventional optical methods. Nevertheless, it is considered that the method used in this investigation provides good control for the determination of average composition, and its success has been demonstrated in Table 1.

Variation in plagioclase composition across the intrusion is shown in Figures 6 and 7 and Table 6. In the early and middle fractionation stages where the compositional trend of plagioclase has been well established, it is shown by a full line in Figure 7, but where in late stages control is less satisfactory this is indicated by a broken line. In the dolerites the average compositions ranges from $An_{63.5}$ to An_{37} where measured, and it becomes progressively enriched in the albite molecule with fractionation. Like olivine and orthopyroxene in the Mg-olivine layer, plagioclase shows a reversal in its fractionation trend; it is richest in the anorthite molecule in the center of the layer. Though average plagioclase compositions only are referred to, plagioclase crystallization in most dolerites was continuous. Small euhedral laths included in ferromagnesian minerals are the most calcic, and are of early crystallization. The bulk of the plagioclase, however, crystallized concurrently with the ferromagnesian minerals, and is commonly subophitically intergrown with them.

The variation in plagioclase composition with fractionation reflects compositional and temperature variations in the magma during crystallization. If this is so, then a line *ab* can be fitted as shown in Figure 7, above which is the orthopyroxene field of crystallization and below which is the field of pigeonite crystallization in the early and middle fractionation stages. Pigeonite also crystallized in the chilled contact dolerite, where it is probably a metastable phase; thus it would appear in the orthopyroxene field of the diagram.

The Fe and Mg contents of the plagioclase are small and show insufficient variation to indicate definite trends, though Mg does appear to decrease with fractionation (Table 15). The values obtained are higher than those reported in plagioclases elsewhere (Ribbe and Smith, 1966), and may be due, as they suggest, to impurities in the plagioclase separated for analysis. Both Ti and Mn occur in trace amounts of less than 0.01 percent, and both show only a very slight increase with fractionation.

Micropegmatite

Micropegmatite comprises a micrographic intergrowth of quartz and alkali feldspar; it occurs throughout the intrusion except in the Mg-olivine layer. It increases in abundance with fractionation. The intergrowth is invariably interstitial as it represents the crystallization of residual liquid. Figure 6 shows the distribution of micropegmatitic residuum through the intrusion. In rapidly cooled phases, its place is commonly taken by a mesostasis with iron oxide inclusions; rarely either quartz or alkali feldspar exist independently.

The alkali feldspar is usually turbid, but where it is fresh it can be identified as a sodic type. In the more acid rocks, zonal plagioclase laths commonly merge into surrounding haloes of micropegmatite (Pl. 8). The proportion of quartz in the micropegmatite varies from 40 to 50 percent. In schlieren, micropegmatite worked along cracks in basic plagioclase in a typical hydrothermal manner.

Opaque Iron Minerals

Opaque iron minerals occur throughout the intrusion, though they are most abundant in dolerite of late fractionation stages where Fe-enrichment reached a maximum. In the Palisades intrusion the opaques are mostly of late crystallization, which implies moderately dry magma conditions. The distribution of total opaques with fractionation can be seen in Table 5 of the micrometric analyses, and in Figure 12.

The main opaques are ilmenite and magnetite, with which minor amounts of ulvo-spinel, pyrite, and chalcopyrite are associated in places, though chalcopyrite also occurs independently. The opaques in most rocks are nearly always titaniferous. No chrome spinel has been identified. The opaque minerals formed skeletal crystals molded on pyroxene, or jagged rods and independent subhedral and anhedral grains associated with interstitial material; grains are evenly distributed in most rocks.

Polished specimens were prepared of dolerites representing stages 2(ii), 5(ii), and 8 of fractionation, and were examined by I. Pontifex for their opaque mineral content. He reports as follows:

In the Mg-olivine layer most of the opaque mineral is ilmenite. Part of some grains consists of titanomagnetite similar to that described below in the pigeonite dolerite. Rarely the ilmenite is in a fine graphic intergrowth with silicate minerals. Chalcopyrite grains, about 0.02 mm diameter, are disseminated through the rock, and in some cases they are associated with the Fe-Ti oxides. Visual observation indicates the following abundances:

Ilmenite	90 percent
Ilmenite-titanomagnetite	7 percent
Chalcopyrite	3 percent.

In the pigeonite dolerite the opaque iron minerals consist mainly of magnetite in anhedral grains up to 2 mm diameter, some of which have inclusions of ilmenite and probable ulvo-spinel. This spinel was identified on its mode of occurrence and limited optical properties (*see* Ramdohr, 1953, p. 677). The ilmenite occurs in lamellae up to 0.005 mm wide. The ulvo-spinel formed bleb-like inclusions about 0.001 mm across, randomly distributed in the magnetite, and commonly localized in aggregates along the margins of the ilmenite lamellae. Some grains of magnetite are free of inclusions but contain the ilmenite lamellae; these are titaniferous magnetite. Grains of chalcopyrite are disseminated through the rock, and in some cases are inclusions in the Fe-Ti oxides. A visual estimate of abundance shows:

Magnetite with ulvo-spinel and ilmenite exsolution	80 percent
Titaniferous magnetite with ilmenite exsolution lamellae	17 percent
Chalcopyrite	3 percent.

The main opaque iron minerals in the ferrodolerite are ilmenite in anhedral irregular grains up to 1 mm across, and titanomagnetite. Chalcopyrite is also present. Some grains of similar size to ilmenite have an extremely fine cloth texture which appears to consist of magnetite with ilmenite intergrowths. The ilmenite bodies in magnetite have a common orientation and measure 0.001 mm by 0.005 mm. The mineral is probably titanomagnetite. Some of the chalcopyrite grains in aggregates up to 0.1 mm across are enclosed by ilmenite. Smaller grains, about 0.03 mm diameter, are disseminated through the rock. Rarely small blebs of Fe-Ti oxides and chalcopyrite formed graphic-like intergrowths with the silicate minerals. A visual estimate indicates the following relative abundances:

Titanomagnetite	50 percent
Ilmenite	48 percent
Chalcopyrite	2 percent.

Other Minerals

Biotite. Biotite is an ubiquitous accessory in the Palisades dolerite. It is most abundant in the Fe-rich dolerites, where it is generally molded on iron oxides or pyroxene, and probably represents a late reaction product of these minerals. With hornblende, it attests to fairly hydrous conditions of crystallization in late fractionation stages.

In the normal dolerite, biotite is golden-brown, and in the acid modifications, it is commonly green-brown. In the Mg-olivine layer it is somewhat more abundant than in the early dolerites and is a rich red-brown, a color which Walker and Poldervaart (1949, p. 646) attribute to chromic oxide in the Karroo picrites, such as that at Elephant's Head.

Biotite is more abundant in the upper than the lower contact dolerite, and in the upper commonly has a poikilitic habit, owing probably to the migration of K-rich solutions. Walker (1940, p. 1075) gives the optical properties for this biotite in the contact dolerite from the location of W-FUC-60, as $\alpha = 1.591$, $\beta = 1.641$, $\gamma = 1.641$ and $2V\ (-) = 0$.

Hornblende. Hornblende developed mainly in the top half of the intrusion and increased in abundance with fractionation. Figure 6 and Table 5 show its distribution in the intrusion. It occurs only in minor amounts in the dolerites, but where fairly hydrous conditions existed during crystallization, as in late-stage fractionation, hornblende is a common dark mineral. It is an important constituent, and may be of primary crystallization in some of the pegmatitic and granophyric dolerites; it commonly occurs in interstitial micropegmatitic areas. Many large grains of green hornblende appear to have crystallized independently; others resulted from reaction of pyroxene; these are free of inclusions so common in the original pyroxene. Some green hornblende contains patches of brown hornblende without distinguishable boundaries between the varieties which are in optical continuity with one another.

Other amphibole grains grade into a blue-green variety whose appearance is reminiscent of sodic types. Uralite formed in those places where the ferromagnesian minerals were affected by hydrothermal alteration or autometamorphism.

Apatite. Apatite formed slender needles that are associated with interstitial micropegmatite. It occurs in most dolerites of the sill, and is particularly common in the granophyric rocks (Pl. 8).

Zircon. Minute zircon crystals can be seen in late-stage fractionation products and these are invariably surrounded by pleochroic haloes in biotite or hornblende.

Sphene. Sphene is rare and occurs in the dolerites of late fractionation in the sill. It may fringe opaque iron minerals, but in the granophyric dolerite an independent well-cleaved variety shows strong brown pleochroism.

Secondary Minerals

Locally in the dolerites a variety of alteration products occurs. Olivine characteristically altered to various serpentine-like minerals including green and golden-brown forms of iddingsite and bowlingite. Chlorite, and rarely calcite, formed in small amounts at grain boundaries of plagioclase and ferromagnesian minerals. Patchy alteration of plagioclase is typically turbid and includes some mica. These alteration products, together with uralite, occur at all levels in the intrusion on minor amounts as patchy alteration to primary minerals. They resulted from the changing hydrothermal conditions and the movement of volatiles during fractionation, and from limited autometamorphic effects.

A secondary mineral of interest is a Si-rich variety of analcite which occurs in the cores and rims of plagioclase laths of some hydrothermally altered vein walls. It is also associated with the late Na-rich veins, where it characteristically formed along cracks in the sodic plagioclase.

Paragenesis and the Evolution of Mineral Series with Fractionation

The interpretation of paragenesis and evolution of the mineral series is necessarily tentative because of the incompleteness of some mineralogical studies. In the basal chilled dolerite, augite, pigeonite, and plagioclase crystallized together with minor amounts of orthopyroxene. Included in the dolerite are microphenocrysts of olivine, augite, bronzite, and plagioclase, which were crystals before the magma was emplaced. The assemblage reflects the metastable conditions of crystallization brought about by rapid cooling of dolerite at the intrusive contact. Pigeonite was preserved intact as a quench product. The paragenetic sequence indicated is that augite and pigeonite crystallized together with plagioclase, and that orthopyroxene, from the reaction of olivine to orthopyroxene, closely followed.

Early and middle fractionation stages. Away from the chilled contact crystallization with fractionation proceeded normally except for the complex mineral relations in the bottom quarter of the intrusion which are believed to have been caused by two-phase intrusion of magma. Introduction of a second phase presumably interrupted the normal course of fractionation initiated by the first, and caused most mineral series to show temporary reversals in their fractionation trends. But, following adjustment to the changed conditions of crystallization, fractional crystallization in the intrusion again proceeded normally.

Augite formed a continuous mineral series with fractionation, and with progressive Fe-enrichment, which occurs mainly at the expense of Mg, it changed towards the mauve-brown ferroaugite. In the Fe-rich fractionation stages both mauve-brown and pale-green ferroaugite developed; the evolution of these is discussed below in the section on the late fractionation stages.

Olivine, orthopyroxene, and pigeonite all formed discontinuous, though in part interrelated, mineral series. With fractionation there is an over-all increase in the Fe/Mg ratio in each series, though reversals in this trend have been observed for olivine and orthopyroxene in the Mg-olivine layer, from which pigeonite is absent.

Resolving the paragenesis and evolution of the layer from the textural relationships between minerals is difficult. The Mg-olivine layer is unique. It should be recalled that compared to the subophitic dolerites above and below, the layer is gabbroic, and poikilitic in part, and that it contains somewhat more biotite. Indeed, the texture is more characteristic of a plutonic than of a hypabyssal basic rock. These features appear to indicate that, in the layer, greater freedom of crystallization existed, owing presumably to a pause in the cooling of the intrusion, during which the magma was quiescent. The central portion is texturally and mineralogically uniform, as if it had crystallized progressively. The basal part appears somewhat heterogeneous, for it contains glomeroporphyritic aggregates of pyroxene and some olivine grains, which appear to have settled from the overlying magma and lodged in it. Thus, there is a concentration of ferromagnesian minerals at the very base of the layer; however, the progressive variation in the abundance of minerals from the bottom to the top of the layer shows that orthopyroxene and biotite do not vary greatly, whereas from the bottom to the center, plagioclase and clinopyroxene decrease and olivine increases, and from the center to the top, these trends are reversed. Another feature of the base of the layer, as mentioned, is the unusual location of exsolution in the orthopyroxene grains. Moreover, olivine is distributed equally between the plagioclase and pyroxene areas of crystallization, though it does not occur in the pyroxene glomeroporphyritic aggregates. The olivine in the pyroxene areas mainly formed the large grains, which tend to be more Fe-rich and corroded than the small.

In the center of the layer, where the olivine is in the plagioclase areas of crystallization, it is commonly fresh and euhedral or subhedral, but where it is in pyroxene, and in particular in orthopyroxene, it has rounded or resorbed margins, and may be partly or completely altered to a serpentine-like mineral. In a bridging situation it has anhedral margins against pyroxene, and subhedral or euhedral against plagioclase (Pl. 5). Of the major mineral components in the layer, plagioclase grains tend to be subhedral, and toward the bottom and top laths commonly are partly or wholly surrounded by the large orthopyroxene grains. Augite grains are mostly anhedral, whereas those of orthopyroxene are always anhedral because they molded on both the plagioclase and augite grains.

From these textural relationships it would appear that crystallization began with olivine and that early plagioclase was in part concurrent with olivine because it incorporated most of the grains in the center of the layer. Plagioclase crystallization would, however, have been continuous during most of the crystallization of the rock, and much of the augite crystallization would overlap with it, as indicated by the subophitic relationship between them in the bottom and top of the layer. Though orthopyroxene was the last major mineral to begin to crystallize, its crystallization also overlapped to some extent with that of the other phases, except olivine. Thus, orthopyroxene is of late formation. In the chilled contact dolerite, it has been shown that olivine may react with magma to form orthopyroxene. Alteration of olivine to serpentine and iron oxides is mostly where olivine occurs in the areas of pyroxene crystallization, and is probably autometamorphic.

The most basic orthopyroxene (En_{83}) is in the early dolerite below the Mg-olivine layer; it probably was introduced as microphenocrysts in the first magma phase. The Bushveld-type orthopyroxene in the layer is much poorer in the En molecule, and its variation in composition through the layer roughly parallels that of olivine, though olivine is consistently more Fe-rich than this co-existing orthopyroxene. The composition of orthopyroxene shows an over-all increase in Fe/Mg ratio with fractionation, and where this ratio is between three-sevenths and two-thirds, the most common form is the Palisades type.

Pigeonite in the Mg-rich fractionation stages has a complementary relationship to orthopyroxene of the Palisades type, which is inverted pigeonite. No evidence has been seen to suggest that pigeonite bears a direct reaction relationship to the olivine of the layer, though pigeonite of primary crystallization does occur above and below the layer. The co-existence of primary pigeonite with orthopyroxene above the layer indicates a complicated history of crystallization for this part of the intrusion, but the full significance of the association cannot be determined until more mineralogical measurement has been made. However, the apparent co-existence of pigeonite with olivine and Bushveld-type orthopyroxene in the early dolerite below the layer is probably a fortuitous association caused by the settling of foreign mineral phases into an environment consistent with pigeonite crystallization.

Late fractionation stages. In the late fractionation stages the magma was enriched in iron, silica, and alkalis. Mineral relations were complicated by the hydrous condition of the magma, which influenced mineral stability fields and controlled, in part, which minerals formed.

As mentioned, the mauve-brown ferroaugite is continuous with the normal augite trend. The evolution of the pale-green variety is more complicated though, for it follows pigeonite in the fractionation sequence, and has a physical association with the mauve-brown ferroaugite which is not unlike that existing between pigeonite and augite in the late-middle fractionation stages. The Fe-rich pigeonite apparently gave way to pale-green ferroaugite in the late stages, but, in addition, some of this ferroaugite formed along the fine exsolution lamellae in the mauve-brown ferroaugite, suggesting that the mauve-brown variety was also involved in the reaction. Petrographic evidence suggests that the ferrohypersthene exsolved from the pale-green ferroaugite (Pl. 7). Possibly this orthopyroxene, in turn, reacted to fayalite and quartz in the very late stages. If reaction relationships exist between these phases, they must be in the direction of pigeonite→ferroaugite→ferrohypersthene→fayalite, because pigeonite and mauve-brown ferroaugite appear before the pale-green ferroaugite in the fractionation sequence, and ferrohypersthene appears before fayalite. Depending, however, on the hydrous condition of the magma, the pale-green ferroaugite may also react to green-brown hornblende. The fractionation process toward more acid types in the Palisades Sill was halted at the ferrodolerite stage by the advance of the crystallization from the top contact.

The two-pyroxene field. Wager and Deer (1939), Muir (1951), and Brown and Vincent (1963) record the presence of two pyroxenes, one brown and the other green, in some of the latest differentiates of the Skaergaard intrusion. Muir, and subsequently Brown and Vincent, showed that both are ferrohedenbergites whose Mg contents are low and decrease with fractionation, that the green variety is somewhat more Fe-rich than the co-existing brown variety, and that both varieties become progressively more Fe-rich with fractionation. Wager and Deer suggested from textural relationships that some of the ferrohedenbergites are an inversion product of iron wollastonite, and Muir, and Brown and Vincent, follow this interpretation in explaining the genesis of some of the green ferrohedenbergites, but Brown and Vincent also recognized the possibility that some are of primary crystallization.

McDougall (1961) also observed two distinct varieties of ferroaugite in the Red Hill intrusion, a pale purple and a pale green, both of which he considered primary phases that crystallized in equilibrium with one another. From a comparison of the optical properties given, it would appear that the green variety is the more Fe-rich of the two. He was unable to explain why the two varieties occurred beyond the two-pyroxene field, but suggested that a small immiscibility gap may appear in this part of the pyroxene system McDougall (1961, p. 677). In a study of the Great Lake Dolerite Sheet, McDougall (1964) recognized that the Ca-rich and Ca-poor pyroxene series may converge toward a continuous solid solution series, with a minimum at the limit of the two-pyroxene field. But he was also aware of an alternative explanation, in that the textural features of the Great Lake intrusion indicate that the Ca-poor pyroxene phase formed during a stage of rapid crystallization, and he suggested that the ferroaugite could be a metastable phase. Poldervaart (1944) also recognized two distinct varieties in the iron-rich dolerite of New Amalfi Sheet.

Poldervaart and Hess (1951, p. 479) have suggested that pigeonite may react with liquid in tholeiitic intrusions to form fayalitic olivine. Muir (1954, p. 384), Brown (1957, p. 525), and Brown and Vincent (1963) concluded, however, that the disappearance of Ca-poor pyroxene is not due to a reaction relationship with fayalitic olivine. Nor could McDougall (1961, p. 679) find any petrographic evidence to suggest that a reaction relationship existed between pigeonite and fayalite, though he observed that pigeonite disappeared when fayalite appeared as a primary phase with fractionation. The data from the Skaergaard and Red Hill intrusions indicate that the trend of the augite series is away from that of the pigeonite trend at the limit of the two-pyroxene field, whereas evidence from the Great Lake intrusion suggests the trend of each converges.

Mineralogical relationships for the early and middle fractionation stages in the Palisades Sill agree largely with those seen in other similar intrusions. Furthermore, the late-stage mineral phases found in the Palisades Sill appear consistent with those reported in other intrusions, but they appear to reveal more clearly the relationships between phases, particularly in the field of the pyroxene quadrilateral beyond the two-pyroxene field. Petrographic evidence suggests that in the Palisades Sill the orthopyroxene-pigeonite trend extends into late fractionation stages where it is represented by a more Ca-rich monoclinic pyroxene, the pale-green ferroaugite, which exsolves(?) orthopyroxene that, in turn, possibly reacts to fayalite. This is not meant to imply, at this point in investigation, that the pyroxene trends of early- and middle-stage fractionation necessarily follow unique trends in the late stages in all intrusions. Physico-chemical conditions vary from intrusion to intrusion, and mineral relations found in one need not necessarily apply to another, as mineral trends in volatile-enriched fractionation stages will be readily modified by slight differences in magma conditions.

Possibly in the Palisades there is a bifurcation of the bronzite-pigeonite trend in the late fractionation stages to a pale-green ferroaugite trend and to a ferrohypersthene-fayalite trend, as the pale-green ferroaugite co-exists with these phases and with the mauve-brown ferroaugite. However, the pale-green ferroaugite presumably can develop from both pigeonite and mauve-brown ferroaugite, and further reaction can result in either orthopyroxene-fayalite or hornblende depending on the hydrous conditions in the magma.

MAJOR ELEMENT CHEMISTRY AND DIFFERENTIATION TRENDS

The chemical study of the Palisades Sill is important because the sill is a complete and readily accessible hypabyssal intrusion of tholeiite, which has undergone extreme fractionation in conditions that, apart from multiple intrusion, may be considered those of an essentially closed chemical system. Multiple intrusion, which is established chemically below, may have been responsible for the accentuation of the fractionation trends and mineral layering. The formation of the Mg-olivine layer, for example, may have resulted in the removal of Mg in early fractionation stages in excess of that normally expected from a magma with a composition of that in the Palisades. Moreover, subsequent progressive fractionation, which took place over a vertical distance of about 1000 feet, proceeded to a more advanced stage than might normally be expected in an intrusion of this thickness: dolerite ranges from hyalosiderite dolerite to fayalite granophyre and granophyric dolerite, and mineral series are complete over a wide compositional range.

The average composition of the basal chilled dolerite is taken as indicating the composition of the original magma. The average composition of the intrusion, however, has been evaluated planimetrically from the whole-rock element distribution plots (Figs. 8, 9, 13-20, 22-24, 27, 28, 31, 33, 34, 36, and 37). These plots show the element content of the whole rock and the mineral phases at each stage in fractionation. They portray element distribution and partitioning in the Englewood Cliff Section, except for Figure 8, which is a composite diagram whose distribution plots have been interpreted from analytical data on rocks of both the Englewood Cliff and the Union City sections. The join of points representing the element values at each stage is taken as the distribution trend of the element with fractionation; each mineral is represented by the same symbol, and abbreviations used are given in Table 7. Full lines are used for joins where values are confidently established, except that the MgO and CaO trends in Figures 8 and 9, and the Ni trend in Figure 38, have been shown by broken lines for clarity in presentation. Elsewhere, most broken lines caution that the analytical

TABLE 7. ABBREVIATIONS USED IN TABLES 10 TO 12 AND FIGURES 12 TO 37

Ol	Olivine
Mg-Ol	Mg-olivine
Fe-Ol	Fe-olivine
Opx	Orthopyroxene
Cpx	Clinopyroxene
Pg	Pigeonite
Pl	Plagioclase
Mt−Il	Magnetic opaque iron minerals (mainly titanomagnetite)
Mt+Il	Magnetite, titaniferous magnetite, and ilmenite
Il	Ilmenite

method used, though reliable, is not the best available for the particular element determination—for example, where major element determinations have been made by optical emission spectroscopy. Broken lines have also been used in the following situations.

(1) For joins between olivine values, because the fayalite analyzed is that from the fayalite granophyre No. 16441 of the Union City Section, and its corresponding location in the Englewood Cliff Section has been interpreted.

(2) For joins between the opaque iron mineral values, as the opaques separated for analysis are probably not entirely representative of the opaques in each fractionation stage. Polished section examination has shown that the opaque iron minerals form complex grains, owing to exsolution, inclusions, and intergrowth of the spinels and sulfides. The opaques are compounded of magnetite, titanomagnetite, ilmenite, ulvo-spinel, chalcopyrite, and so on, and it would be impossible to make a separation suitable for a pure mineral analysis. Magnetite and ilmenite, in particular, are intimately associated. The opaque iron minerals analyzed are the most magnetic fraction of a concentrate of opaques, so that results are probably biased toward values for magnetite.

(3) For the join of some Cr values in clinopyroxenes, as the sum of the Cr content of the minerals, after appropriate modal corrections, far exceeds the actual Cr content of the whole rock. There are two possible explanations for this. First, standard control of working curves for high Cr values was inadequate. Repeat analyses by different methods, however, only confirmed the high values obtained. More acceptable, therefore, is the second possibility,

that minute inclusions (including possibly turbid alteration) in the pyroxenes are mineral phases with abnormally high Cr content. Thus, element values obtained for some clinopyroxenes may not be those for the pure mineral, but include contamination by the inclusions.

The joins between element values of minerals W-N-60 and W-E-60, and joins between element values of whole rocks W-J-60 and W-E-60 (Figs. 13-20, 22-24, 27, 28, 31, 33, 34, 36, and 37) have been made for the convenience of discussing element partitioning between minerals and the whole rocks, though the distribution trends are incomplete for those elements which concentrated in late fractionation stages. Had the fayalite granophyre and the granophyric dolerite been included in the whole-rock and mineral distribution trends, they would have modified them somewhat. Both are part of the fractionation series, but attempts to separate wanted minerals were unsuccessful except for the fayalite of the fayalite granophyre. Element content in the granophyric dolerite is shown by point X in the distribution plots Figs. 9, 3-20, 22-24, 27, 28, 31, 33, 34, 36, and 37). By excluding these dolerites from joins between fractionation stages, the distribution trends of the rocks and minerals can be compared directly.

Silicate Analyses

The silicate analyses and norms of the dolerites representing the various stages in fractionation are given in Table 8. Sixteen new analyses are presented. They are mainly of rocks from the Englewood Cliff Section, and have been arranged in sequence according to height above the base of the intrusion, a sequence that corresponds to the order of progressive crystallization, except for some late-stage dolerites. Representative dolerites of the fayalite granophyre stage are from the Union City Section. The remaining two analyses, 14-105 and 14-109½, draw attention to the closeness of bulk chemical composition of the pigeonite and bronzite dolerites immediately below and above the internal chilled contact at Haverstraw.

Fractionation Indices

It is common practice to use a fractionation index to arrange rocks according to the sequence of fractionation. This procedure has the advantage that widespread rocks in a province can be compared, and provides a common control for comparative studies between provinces.

In the case of tholeiitic intrusions, the mafic index

$$\frac{(FeO + Fe_2O_3) \times 100}{FeO + Fe_2O_3 + MgO},$$

Table 8. Chemical Analyses and Norms

ENGLEWOOD CLIFF SECTION

Specimen	W-889LC-60 (a)	W-865-60	W-824-60	W-804-60	W-U-60	W-R-60	W-N-60	W-J-6
Height*	1	30	70	90	215	365	560	68
SiO_2	51.98	51.81	47.41	52.55	52.20	52.24	52.05	52.6
TiO_2	1.21	1.16	0.89	1.07	1.02	1.15	1.24	2.7
Al_2O_3	14.48	13.36	8.66	12.53	14.23	16.04	16.59	11.9
Fe_2O_3	1.37	2.24	2.81	2.20	1.68	2.35	2.41	3.9
FeO	8.92	8.44	11.15	8.01	7.95	7.69	7.70	11.0
MnO	0.16	0.17	0.20	0.15	0.16	0.16	0.15	0.1
MgO	7.59	9.25	19.29	9.58	8.50	5.80	5.08	3.9
CaO	10.33	11.12	6.76	10.83	11.44	10.61	9.80	8.0
Na_2O	2.04	1.98	1.35	1.94	2.07	2.50	2.96	2.7
K_2O	0.84	0.59	0.43	0.46	0.46	0.58	0.81	1.2
P_2O_5	0.14	0.14	0.10	0.12	0.11	0.15	0.20	0.3
H_2O+	0.88	0.25	1.45	0.88	0.68	1.10	1.25	1.6
H_2O-	0.16	0.23	0.11	0.20	0.19	0.19	0.27	0.2
CO_2								n
	100.10	100.74	100.61	100.52	100.69	100.54	100.51	100.14
Analyst		Y.Chiba	Y.Chiba	Y.Chiba	Y.Chiba	Y.Chiba	Y.Chiba	H.B.Wi
Mafic index	57.55	53.58	41.98	51.59	53.11	63.38	66.55	79.1
Felsic index	21.80	18.77	20.84	18.14	18.11	22.49	27.78	33.5
Total Fe as FeO	10.15	10.46	13.68	9.99	9.46	9.81	9.87	14.5

NORMS

Q	2.80	1.61		3.44	2.29	4.41	2.63	8.0
Or	4.73	3.28	2.39	2.56	2.56	3.23	5.06	7.6
Ab	16.92	16.66	11.53	16.56	17.50	20.96	24.94	23.5
An	28.23	25.82	16.21	24.27	28.22	31.03	29.53	16.2
Wo	9.28	11.85	6.84	11.95	11.55	8.62	7.60	9.0
Di En	5.19	7.33	4.66	6.76	7.03	4.79	4.05	3.9
Fs	3.72	3.82	1.64	4.67	3.87	3.48	3.30	5.1
En	13.74	15.80	19.17	17.19	14.22	9.71	8.65	5.9
Hy Fs	9.86	8.22	6.74	6.61	7.84	7.07	7.07	7.7
Ol Fo			17.08					
Fa			6.69					
Mt	2.03	3.25	4.18	3.25	2.50	3.37	3.50	5.6
Il	2.28	2.22	1.65	2.09	2.01	2.21	2.34	5.2
Ap	0.24	0.34	0.34	0.34	0.34	0.34	0.34	0.7
H_2O	1.04	0.48	1.56	1.08	0.87	1.29	1.52	1.2
	100.05	100.68	100.68	100.77	100.80	100.52	100.53	100.1

(a) Average analysis nd Not detected * Height above base in feet

TABLE 8. (CONTINUED)

LEWOOD CLIFF SECTION				UNION CITY SECTION				HAVERSTRAW	
-60	W-E-60	W-D-60	W-FUC-60 (a)	W-WU5-61	16441+	W-WU3-61	W-WU17b-61	14-105	14-109½
90	805	830	900	960	≈980	990		102	110
59	50.10	50.38	51.5	54.96	53.6	57.94	52.20	50.18	51.03
56	3.57	3.37	1.2	2.74	2.15	1.81	1.24	1.57	1.56
26	10.91	11.28	14.9	11.93	11.2	11.31	13.74	14.63	14.35
00	4.75	4.77	1.7	2.84	19.2as	2.44	1.58	2.58	2.60
80	14.00	13.17	8.1	12.59	Fe_2O_3	11.86	10.32	7.77	7.92
19	0.25	0.21	0.15	0.21	0.28	0.20	0.21	0.14	0.16
35	3.79	3.44	7.0	2.29	1.80	2.18	7.55	8.04	7.60
66	8.01	6.27	10.0	6.75	6.15	5.65	10.60	10.29	10.23
68	2.37	2.44	2.4	3.07	2.70	3.11	2.64	2.05	1.93
52	1.22	1.39	1.0	1.55	1.98	2.23	0.28	0.60	0.63
85	0.26	0.28	0.2	0.36	1.01	0.72	0.19	0.25	0.21
52	0.90	2.17	1.2	1.10		1.12	tr	1.11	0.82
43	0.29	0.23	0.5	0.17		0.10	0.27	0.55	0.39
d		0.11						nd	nd
41	100.42	99.51	99.9	100.56	100.07	100.67	100.82	99.76	99.43
Wiik	Y.Chiba	H.B.Wiik		Y.Chiba	C.D.Branch	Y.Chiba	Y.Chiba	H.B.Wiik	H.B.Wiik
08	83.18	83.91	58.33	87.07		86.77	61.18	56.28	58.05
88	30.94	37.92	25.37	40.63	43.21	48.58	21.59	20.47	20.01
50	18.28	17.46	9.6	15.15	17.28	14.06	11.74	10.09	10.26
				NORMS					
33	6.83	9.16	1.36	10.46		13.10	0.90	2.59	4.61
12	7.34	8.21	6.12	9.18		12.96	1.56	3.55	3.72
22	20.13	20.65	20.44	25.89		26.30	22.43	17.35	16.33
61	15.43	15.72	26.69	14.29		10.45	24.68	28.95	28.63
48	9.56	5.66	9.25	7.02		5.36	10.89	8.55	8.66
49	3.58	2.09	5.28	1.90		1.36	7.06	5.38	5.32
18	6.15	3.68	2.24	5.48		4.30	3.08	2.64	2.85
89	5.90	6.47	12.22	3.83		4.09	11.82	14.64	13.60
89	10.12	11.39	9.52	11.10		12.93	12.91	7.17	7.27
41	6.85	6.92	2.55	4.18		3.25	2.27	3.74	3.77
98	6.80	6.40	2.28	5.23		3.52	2.34	2.98	2.96
02	0.57	0.66	0.34	0.87		1.75	0.64	0.59	0.50
95	1.19	2.40	1.77	1.27		1.22	0.27	1.66	1.21
57	100.45	99.42	100.0	100.70		100.59	100.85	99.77	99.44

ay fluorescence analysis

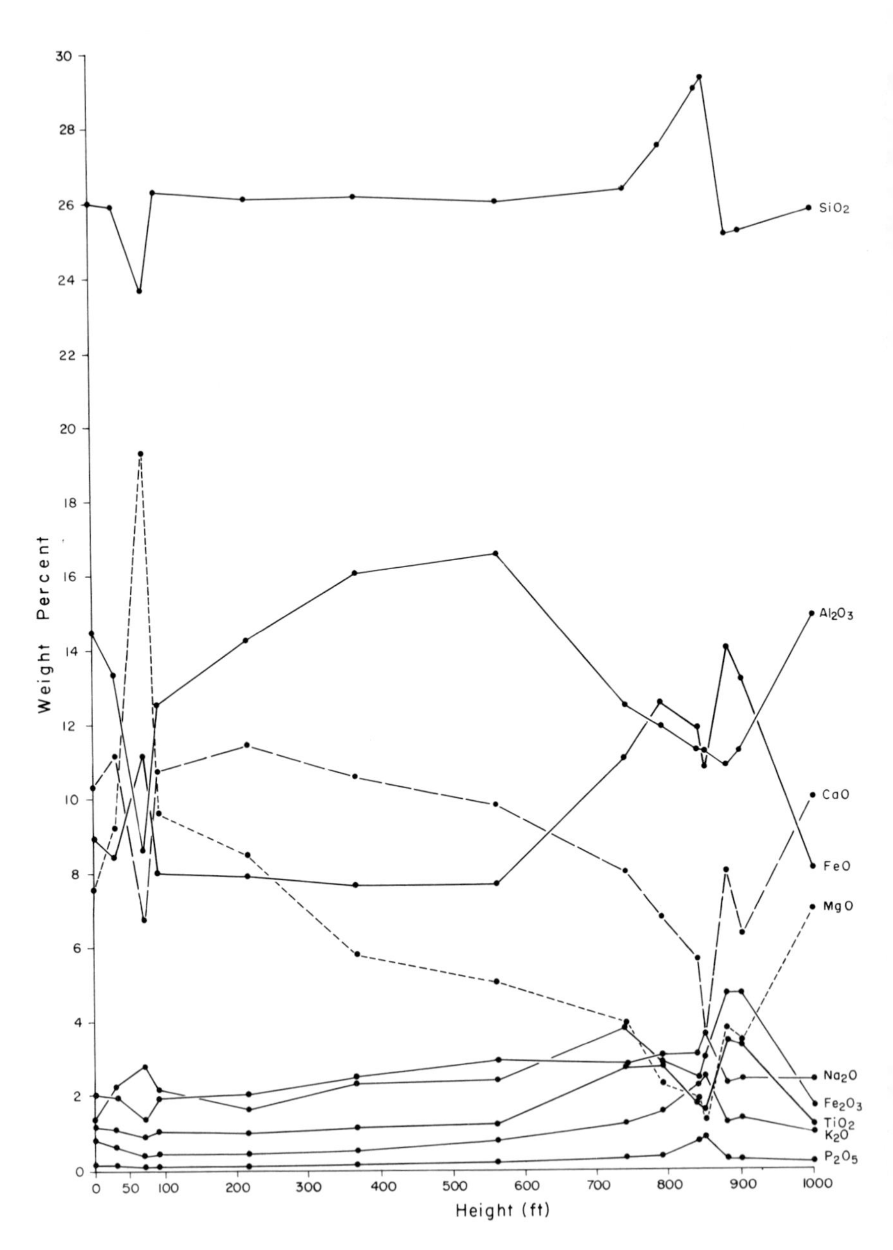

Figure 8. Distribution trends of the major oxides, a composite diagram of the Englewood Cliff and Union City Sections. Silica shown at half scale.

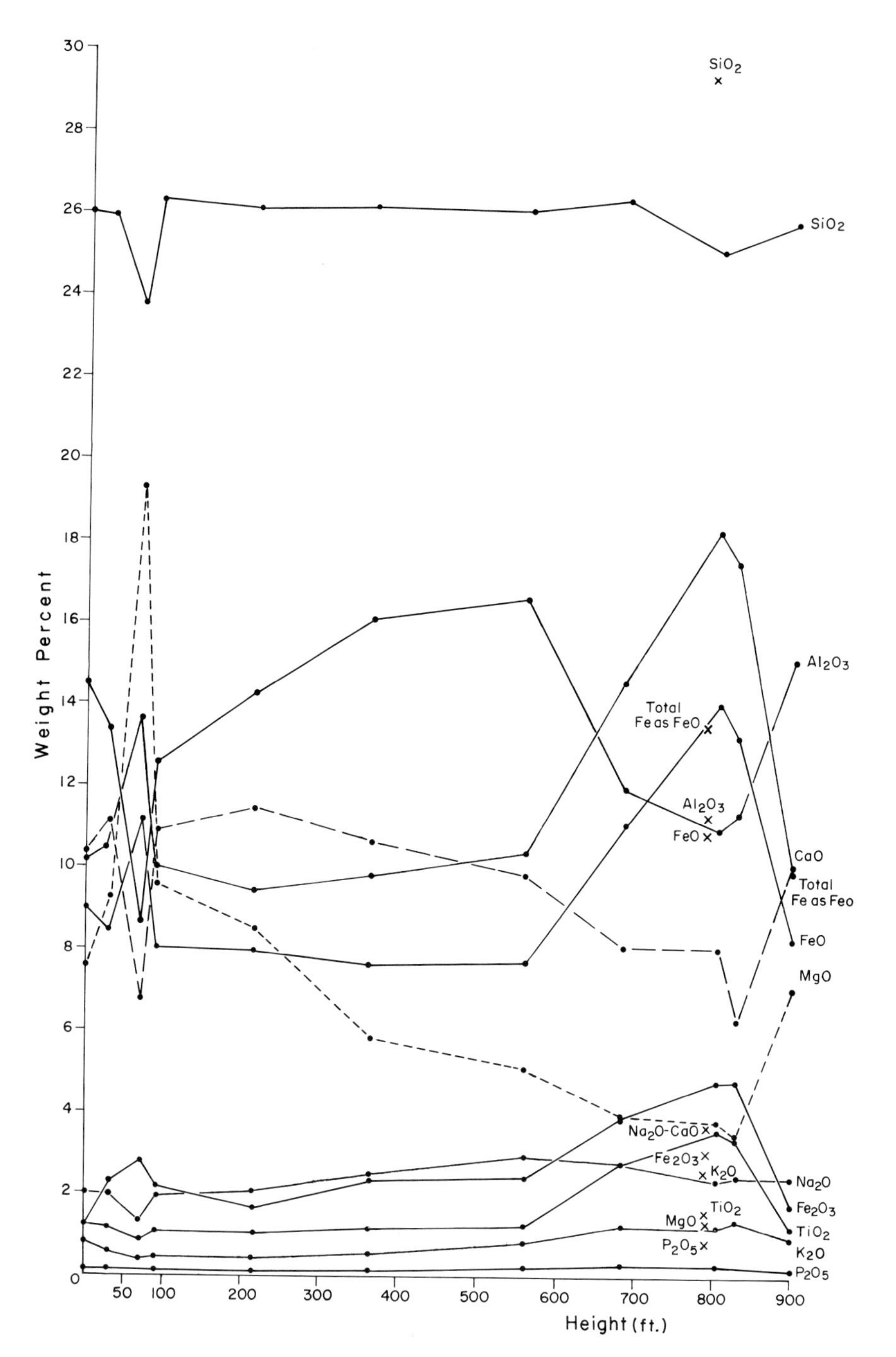

Figure 9. Distribution trends of the major oxides, Englewood Cliff Section. Silica shown at half scale.

and less commonly the felsic index

$$\frac{(Na_2O + K_2O) \times 100}{Na_2O + K_2O + CaO},$$

have been widely used. They identify the two main trends of differentiation in tholeiitic intrusions. For the convenience of comparing the Palisades with other similar intrusions, these indices have been included in Table 8. Plots of the indices against the height of specimens above the base of the intrusion would show that they closely parallel each other, except that the plot of the mafic index would emphasize the variations with fractionation more in the early stages, and less in the late stages.

The mafic index for the Palisades dolerites ranges from 42 to 91. This compares with a range of about 41 to 96 for all Tasmanian Mesozoic dolerites (Edwards, 1942, and McDougall, 1962, 1964) and of about 26 to 96 for all dolerites in the Karroo province (Walker and Poldervaart, 1949). Thus fractionation within the Palisades Sill covers a wide range, particularly when considered in relation to its size.

No index, however, includes all the parameters that control magma behavior with fractionation. An intrusion such as the Palisades can be considered to be a complete natural chemical system, and the changes that occur with progressive fractionation can be defined only by accounting for all chemical variations within it. Because of the obvious deficiencies of indices in defining the complete differentiation process and because the Palisades Sill is a complete intrusion that can be systematically sampled, height above the base has been used for control against which rocks are measured. Variations in height correspond to a factor which defines the degree of fractionation shown by each rock. This procedure does present some problems in dealing with the very late-stage fractionation products, but it has the advantage that variations in magma behavior with fractionation can be viewed as a whole or as separate phases as desired.

Furthermore, analyses used in chemical discussion have not been recalculated anhydrous, as water is an integral part of the chemical system and does not vary greatly between the rocks chosen for study. In these water content mostly ranges from 1 to 1.5 percent. Recalculation of analyses as anhydrous would change values only slightly, and would have little effect on comparative studies with other intrusions.

Differentiation Trends

Both old and new analyses of the Palisades dolerites have been plotted in the MgO-FeO—($Na_2O + K_2O$) and the MgO—(Total Fe as FeO)—

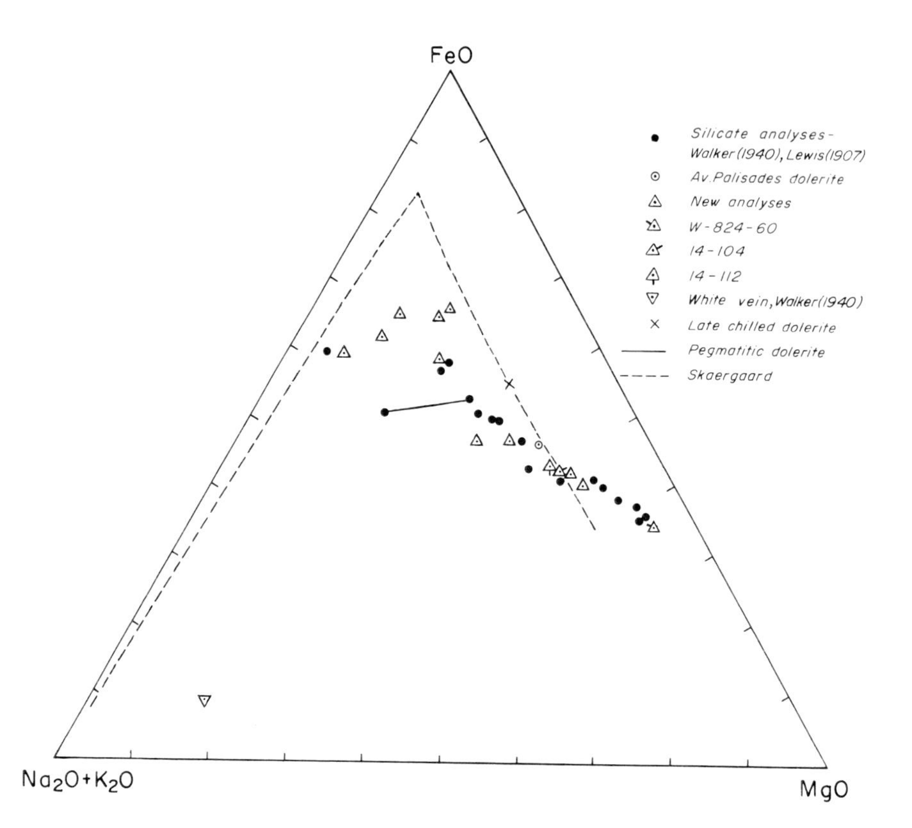

Figure 10. MgO-FeO-(Na_2O + K_2O) diagram showing the differentiation trend in the Palisades Sill.

(Na_2O + K_2O) triangular diagrams (Figs. 10 and 11). Each shows the differentiation trend of the Palisades intrusion, together with that of the Skaergaard, which has been included for comparison. The diagrams indicate that the Palisades intrusion fractionated from early to late stages toward Fe-enrichment with only slight alkali-enrichment, and that in the very late stages the trend changed and moved exclusively toward alkali-enrichment. The analyses (Table 8 and Fig. 8) also show that silica-enrichment was concurrent with alkali-enrichment.

In the course of comparing differentiation trends and average compositions of tholeiitic hypabyssal intrusions throughout the world, Walker and Poldervaart (1949) and Hess (1960) included the Palisades Sill and showed that its trend is similar. Since their work more data have become available on some of the provinces, and, in particular, on the Tasmanian province

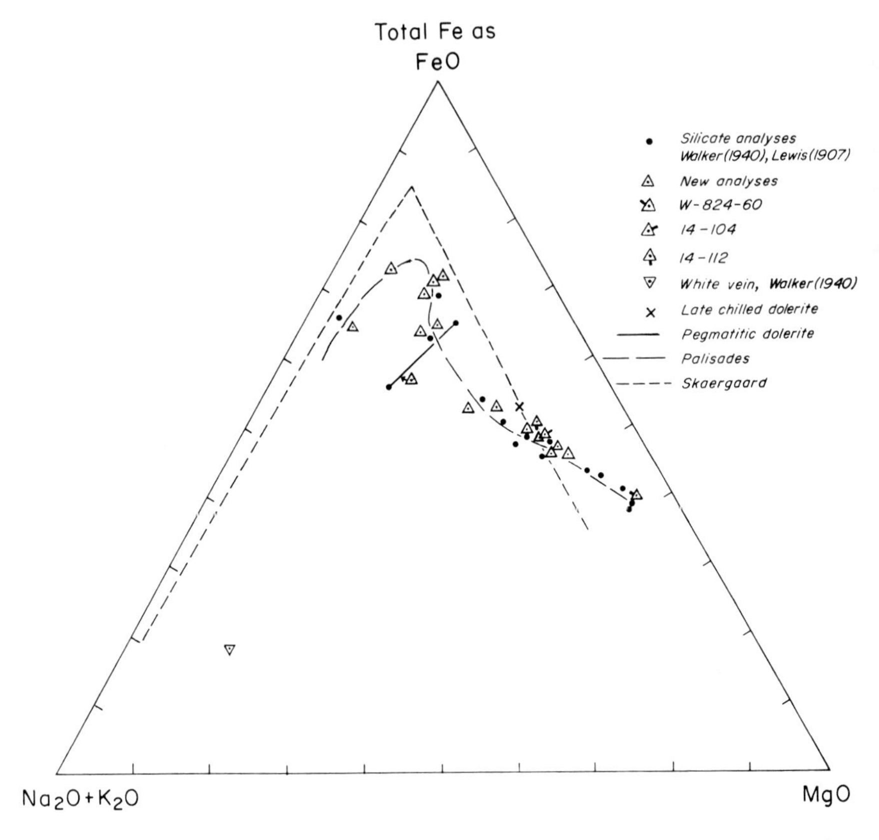

Figure 11. MgO-(total Fe as FeO)-(Na_2O + K_2O) diagram showing the differentiation trend in the Palisades Sill.

(McDougall, 1962, 1964) and on the Antarctic province (Gunn, 1962, 1963, 1966; Hamilton, 1965). The new information has not changed basically the form of the established trends, apart from extending them slightly in a predictable manner.

Composition of the Original Magma

The Palisades dolerite is a typical tholeiite similar in composition to the other Mesozoic tholeiites that occur elsewhere in the world. Table 9 presents a comparison between the composition of the basal chilled Palisades dolerite and the average chilled contact dolerites of the Karroo and Tasmanian provinces and the average tholeiite compiled by Nockolds (1954). The Karroo average is analysis A (Walker and Poldervaart, 1949, Table 16), an

TABLE 9. AVERAGE COMPOSITION OF ORIGINAL MAGMA AND THE PALISADES INTRUSION

	1	2	3	4	5	6	7
SiO_2	50.8	53.2	51.4	52.0	51.8	52.3	52.35
TiO_2	2.05	0.65	1.1	1.2	1.65	1.55	1.6
Al_2O_3	14.1	15.4	15.3	14.5	13.95	13.95	14.0
Fe_2O_3	2.9	0.75	1.0	1.35	2.75	2.65	2.65
FeO	9.1	8.3	9.6	8.9	9.2	9.0	9.2
MnO	0.2	0.15	0.3	0.15	0.15	0.15	0.15
MgO	6.3	6.7	8.1	7.6	6.5	6.35	6.15
CaO	10.4	11.0	9.6	10.3	9.65	9.4	9.35
Na_2O	2.25	1.65	1.8	2.0	2.4	2.5	2.5
K_2O	0.8	1.05	0.7	0.85	0.75	0.85	0.85
P_2O_5	0.25	0.1	0.1	0.15	0.2	0.2	0.25
H_2O	0.9	1.1	1.0	1.05	1.2	1.3	1.3
	100.05	100.05	100.00	100.05	100.2	100.2	100.35
NORMS							
Q	3.6	5.0	2.4	2.8	3.95	4.4	4.6
Or	5.0	6.1	3.9	4.7	4.65	5.35	5.4
Ab	18.85	13.9	15.2	16.9	20.45	20.85	21.0
An	25.85	31.5	31.7	28.2	24.75	24.2	24.2
Di Wo	10.2	9.2	6.2	9.3	9.3	9.05	8.7
Di En	5.7	4.9	3.4	5.2	5.1	4.8	4.7
Di Fs	4.1	4.0	2.6	3.7	3.8	3.95	3.7
Hy En	10.05	11.9	16.85	13.75	11.2	11.1	10.6
Hy Fs	7.2	9.8	12.85	9.85	8.3	8.05	8.7
Mt	4.2	1.0	1.4	2.0	4.1	3.8	3.8
Il	3.9	1.3	2.15	2.3	3.1	3.0	3.0
Ap	0.5	0.3	0.35	0.25	0.15	0.15	0.5
H_2O	0.9	1.1	1.0	1.05	1.25	1.3	1.3
	100.05	100.0	100.0	100.0	100.1	100.0	100.2

(1) Average tholeiitic basalt and dolerite, Nockolds (1954, p. 1021)
(2) Average Tasmanian chilled dolerite, McDougall (1962, Table 7)
(3) Average Karroo chilled dolerite, Walker and Poldervaart (1949, Table 16, Anal. A), readjusted to include 1 percent H_2O
(4) Average Palisades basal-chilled dolerite
(5) Average composition of the Palisades intrusion, Englewood Cliff Section, a planimetric estimate from 0 to 900 feet, excluding W-F-60, Figure 11
(6) As for 5, but including W-F-60 element values
(7) Average composition of the Palisades intrusion, composite section, a planimetric estimate from 0 to 1000 feet, Figure 10

average of 6 analyses readjusted for 1 percent H_2O. It may or may not be representative of the entire province. The Tasmanian average (McDougall, 1962, Table 7) represents 13 analyses, and the average tholeiite of Nockolds (1954, p. 1021), 137 analyses.

The average composition of the original Palisades magma is given also under specimen W-889LC-60 in Tables 8 and 10. This analysis is an average of 7 basal chilled dolerites. Six of the rocks come from the chilled selvage of the basal contact over a distance of about 10 miles from Englewood Cliff to Kings Bluff, and the seventh is from Kingston, New Jersey, about 40 miles southeast of Kings Bluff. All these rocks are almost identical in composition, and indicate that the original magma was remarkably uniform in composition upon emplacement. The analysis given under W-FUC-60 is an average of four chilled dolerites from the top contact whose locations are as widespread as those of the basal contact. However, they show somewhat greater variation in composition, which apparently reflects the variation in hydrothermal alteration along the top contact. Differences are mainly in alkalis, CaO, and water content, and in the oxidation state of Fe.

Composition of the Intrusion

The average composition of the whole intrusion is given in Tables 9 and 10, where it can be compared with the composition of the original magma as represented by the basal chilled contact. Four alternative average compositions of the intrusion have been computed. It was necessary to know whether a limited average computed by excluding some of the late fractionation products provided an adequate average for the study of element partitioning between minerals, and for equating the sum of the element contents of the minerals to the over-all element composition of the intrusion, and thus whether the comparative study could be made from data on the Englewood Cliff Section only.

Planimetric estimates have been computed for (i) the Englewood Cliff Section from Figure 9 over the range 0 to 900 feet, including and excluding element values for rock W-F-60, and over the range 30 to 805 feet excluding W-F-60; and for (ii) the composite section of the intrusion from Figure 8 over the range 0 to 1000 feet including element values of all late-stage fractionation products. It can be seen in Table 9 and 10 that there is little difference in average compositions from each procedure followed, and that the alternative using the 0 to 900-foot range excluding W-F-60 for the Englewood Cliff Section provides a satisfactory working average for comparative studies. Moreover, little inaccuracy results if the range 30 to 805 feet is used; differences in average element values are small relative to the differences between the average element composition of the intrusion and that of the original magma. This is because the few excluded very late-stage products occur over a small vertical range relative to the size of the intrusion.

The main variations between the element contents of the intrusion computed from the composite section and those of the 0 to 900-foot range of the Englewood Cliff Section (excluding some late-stage products) are in SiO_2, Na_2O and K_2O, and P_2O_5 which increase 0.5, 0.1, and 0.05 percent respectively in the composite average, and in MgO and CaO, which decrease 0.4 and 0.3 percent respectively.

The average trace element content of the intrusion has also been estimated planimetrically on the basis of the 0 to 900- and the 30 to 805-foot ranges in the Englewood Cliff Section, and again exclude the contribution to values made by rock W-F-60 and some other late-stage fractionation products. The limitation on the measurement of the element content of some minerals made this procedure necessary, so that the element content of the minerals can be compared directly to that of the rocks. Examination of the trace element distribution plots shows that if the element values of W-F-60 (point X in the plots) were included in the averages together with values given in Tables 8 and 13 for the other excluded late-stage products, they would not vary the average for Cr, Ga, Ni, Sc, and Sr, whereas the average values used for Ba, Cu, La, Y, and Zr are slightly low, and those for Co and V are slightly high.

The differences in the composition of the original magma and the average composition of the intrusion shown in Table 10 result from the multiple injection of magmas of somewhat different composition. The second magma phase differed from the first in the concentrations of most elements. The differences are too large, consistent, and predictable to attribute them to analytical inaccuracy. The second phase was richer in Fe, Ti, Ba, Cu, Sr, V, and Zr, all elements that concentrate in late fractionation stages; it was poorer in Mg, Ca, Cr, and Ni, which concentrate mainly in the early stages. As expected, elements with small concentration ranges in the intrusion—Mn, Co, La, Sc, and Y—show the least variation between the phases. The second magma phase obviously represents a somewhat later fractionation stage than the original magma.

Element Partitioning between Minerals in Rocks and the Intrusion

An attempt has been made to equate the sum of the element content of minerals to the element content of each rock, and to the average element content of the intrusion. This is a testing appraisal of the analytical results and serves as a useful guide on element partitioning, indicating where elements are located in a rock. It must be emphasized, however, that the results obtained are tentative, as they include estimates and approximations which can greatly influence them.

TABLE 10. ELEMENT PARTITIONING IN INTRUSION

	Fe (%)	Mg (%)	Ca (%)	Mn (%)	Ti (%)	Al (%)	Ba (ppm)	Co (ppm)	Cr (ppm)
AVERAGE COMPOSITION OF CHILLED DOLERITE									
Base									
W-889LC-60	7.9	4.58	7.39	0.12	0.73	7.66	195	53	315
Top									
W-FUC-60	7.5	4.2	7.2	0.12	0.72	7.65	165	52	270
AVERAGE COMPOSITION OF INTRUSION									
0-900 feet*	9.05	3.9	6.9	0.12	1	7.4	216	56	189
30-805 feet*	8.8	4.0	7.05	0.12	0.95	7.45	209	56	190
COMPOSITION OF MINERALS (based on modal values)									
Σ Minerals									
30-805 feet	7.6	3.48	7.44	0.11	0.78			46	308
Mg − O1	0.09	0.11		0.002	tr	tr		2	
Fe − O1	0.25	0.01		0.003	0.002	tr		1	
Opx	1.07	0.90	0.13	0.016	0.014	0.05		8	28
Cpx	3.45	2.43	3.47	0.075	0.114	0.35		29	253
Pl	0.31	0.03	3.84				123		3
Mt − I1	1.9			0.016	0.2			6	24
Il	0.53				0.45				
COMPOSITION OF MINERALS (based on normative and modal values)									
Σ Minerals									
30-805 feet	7.26	3.22	7.44	0.10	0.77			43	283
Mg − O1	0.09	0.11		0.002	tr	tr		2	
Fe − O1	0.25	0.01		0.003	0.002	tr		1	
Opx	1.03	0.87	0.13	0.015	0.014	0.05		8	27
Cpx	3.12	2.19	3.14	0.068	0.103	0.32		26	229
Pl	0.34	0.04	4.17				134		3
Mt − I1	1.9			0.016	0.2			6	24
Il	0.53				0.45				

* Estimated planimetrically from Figures 11, 13-20, 22-24, 27, 28, 31, 33, 34, 36, and 37 excluding W-F-60 (x) values

M Average modal value for intrusion vertical thickness 30 — 805 feet, *see* Fig. 12)

Table 10. (Continued)

Cu (ppm)	Ga (ppm)	La (ppm)	Ni (ppm)	Sc (ppm)	Sr (ppm)	V (ppm)	Y (ppm)	Zr (ppm)	Average Mode (M) Norm (N) (Wt. %)
AVERAGE COMPOSITION OF CHILLED DOLERITE									
110	14	30	95	37	175	235	29	120	
115		14	110	36	150	260	24	105	
AVERAGE COMPOSITION OF INTRUSION									
135	17	27	84	36	187	315	30	129	
130	17	26	85	35	191	301	29	126	
COMPOSITION OF MINERALS (based on modal values)									
			101	35	157	378			
tr		tr	6	tr		tr	tr		0.46(M)
tr		tr	tr	tr		tr	tr		0.49(M)
1		tr	25	3		11	1		6.27(M)
8		6	59	32	5	160	9		31.18(M)
7	8				152	4	4		46.34(M)
5	2	4	11	tr		203	2	8	3.0(M)
									1.5(M)
COMPOSITION OF MINERALS (based on normative and modal values)									
			95	32	169	363			
tr		tr	6	tr		tr	tr		0.46(M)
tr		tr	tr	tr		tr	tr		0.49(M)
1		tr	24	3		11	1		6.04(N)
7		5	54	29	4	145	8		28.20(N)
8	9				165	4	5		50.28(N)
5	2	4	11	tr		203	2	8	3.0(M)
									1.5(M)

N Average normative value for intrusion (vertical thickness 30—805 feet, *see* Fig. 12)
tr Trace

Two rocks representing the early, and one each from the middle and late fractionation stages, have been used in the individual rock study, the results of which are presented in Table 12. The assessments have been based on modal values, but where, as in the two early-stage rocks W-824-60 and W-U-60, the normative and modal minerals correspond, values for the former are given also. Moreover, W-824-60 and W-U-60 have yielded most mineralogical data; indeed W-824-60 is the only rock studied that is so fresh and of such texture that each mineral can be separated in a virtually pure form.

TABLE 11. ESTIMATED MINERAL COMPOSITIONS AND MODES USED IN ELEMENT PARTITIONING COMPUTATIONS

	Fe (%)	Mg (%)	Ca (%)	Mn (%)	Ti (%)	Modes (Wt. %)	Opaque Minerals
W-824-60							
Ol (Fo_{73})	19.04	22.3					
Opx (En_{72})	14.37	15.98	1.43				
Mt–Il	68				2.2	0.3	Titanomagnetite
Il	35				30	2.7	Ilmenite
W-U-60							
Opx (En_{60})	19.8	12.7	2.86				
Mt–Il	64				5.7	0.5	Titanomagnetite and Magnetite
Il	35				30	1.5	Ilmenite
W-N-60							
Pg	25	7	2.5	0.5	0.5		
	62.5				7.5	3.0	Magnetite
Mt+Il						2.0	Titaniferous magnetite
W-E-60							
Ol (Fo_{10})	51.3	2.4					
Mt+Il	62.5				7.6	5.0	Titanomagnetite
Il	35				30	5.0	Ilmenite

The computations of the results include the following approximations. Mineral densities were estimated from plots relating density to mineral composition (Deer and others, 1962, 1963). Modal values were calculated from micrometric data (Table 5), and these are liable to error where mineral abundances are small; for example, the opaque iron minerals and apatite—which strongly influence the distributions of some elements. The major element composition of pigeonite in rock W-N-60 has been estimated relative to that of the co-existing augite, and is given in Table 11. To obtain trace element values for this pigeonite, it has been assumed that variations in pigeonite composition with fractionation are continuous with the Palisades-type orthopyroxene, and values have been calculated on a proportional basis relative to the compositions of the co-existing augites in rocks W-N-60 and W-U-60. The predicted trace composition of the pigeonite is 170 ppm Co, 15 ppm Cr, 20 ppm La, 170 ppm Ni, 75 ppm Sc, 300 ppm V, and 30 ppm Y. Values used for Fe-olivine in rock W-E-60 are those determined in the olivine of rock 16441. No correction has been attempted for contamination indicated in some minerals (*see* Table 11).

In the assessments of element partitioning in individual rocks and in the intrusion, not all mineral phases have been accounted for. Unaccounted minerals would have little effect on the assessments of the early-stage rocks, but they would make an important contribution to the element content of the late-stage rocks. Moreover, the exclusion of hornblende, biotite, micropegmatite, apatite, and the alteration products, from the total element content, affects the assessment of element partitioning for the whole intrusion, as these minerals constitute about 10 percent by weight of the intrusion.

Figure 12 shows the distribution trends of modal (M) and normative (N) minerals in the intrusion against height above the base. The diagram has been used for a planimetric estimate of the average modal and normative mineral contents of the intrusion over the range 30 to 805 feet in the Englewood Cliff Section, and in compiling it certain interpretations were necessary.

(1) The modal content of olivine in the Mg-olivine layer has been assumed to be equivalent to 24 percent by weight over a vertical thickness of 15 feet, and the element values for this mineral in the intrusion have been computed by using the values obtained for the olivine in rock W-824-60.

(2) The contribution of orthopyroxene to element values has been limited to the range from 50 to 250 feet, and the values used are an average of those for rocks W-824-60 and W-U-60.

(3) The modal content of olivine in the Fe-olivine horizon has been assumed to be 2.5 percent by weight over a vertical thickness of 150 feet. This is somewhat greater than the actual Fe-olivine range in the intrusion. Extension of the range was made to compensate for the Fe-rich orthopyroxene as

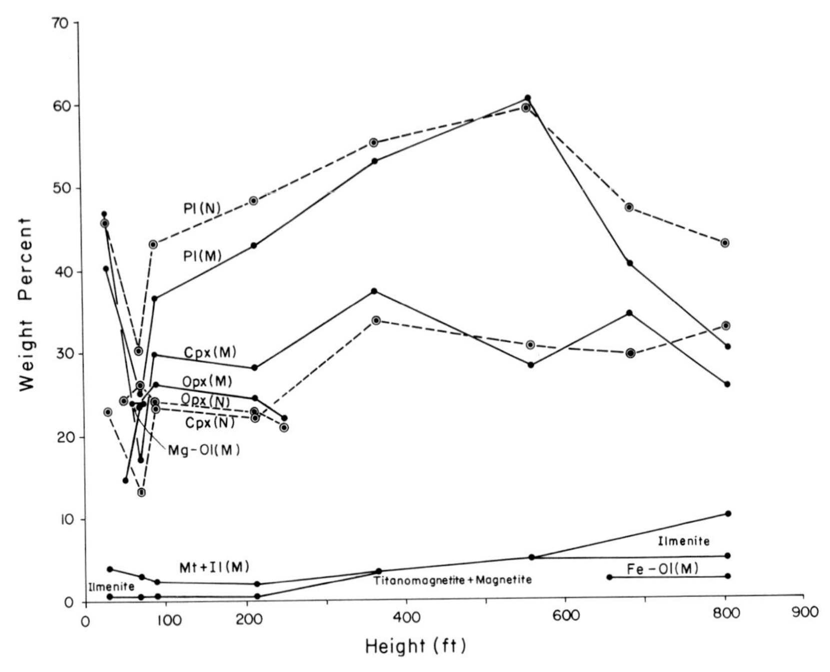

Figure 12. Distribution trends of the modal (M) and normative (N) minerals, Englewood Cliff Section.

little is known about the element content of this mineral at present. Thus, the element content of the fayalite has been used to calculate what is probably only a fairly rough estimate of the contribution made by the orthorhombic mineral phases to the composition of the intrusion in the late fractionation stages. Iron-olivine, however, does not appear in the norm, and in the assessment based on norms (Table 10 and 12) the modal values have been used for both the olivine and the opaque iron mineral calculations. This necessitated subtracting an amount equivalent to Fe-olivine from the normative pyroxene of rocks W-J-60 and W-E-60. Elsewhere, in rocks whose normative pyroxenes do not correspond to the modal, the total normative pyroxene (Di + Hy) has been taken as the clinopyroxene value.

(4) As mentioned previously, the opaque iron minerals analyzed do not represent all the opaques in each rock, and element values are probably biased towards those of magnetite and titanomagnetite. Polished-section examination established that much ilmenite occurs in some rocks, particularly those of early and late fractionation stages. An attempt has been made, therefore, to delineate the ilmenite separately from the magnetite and titano-

magnetite in Figure 12. This subdivision has been made on the basis of the proportion of each opaque identified in the mineragraphic examinations, and on an assessment of the relative amounts of magnetic and nonmagnetic portions obtained in the mineral separations from each rock. It is assumed that, because the opaques analyzed were the most magnetic fraction, their modal proportion is represented by the magnetite + titanomagnetite field in the diagram. Iron has not been determined in them, and has been estimated on the basis that an average magnetite contains about 70 percent total Fe from which the determined Ti content has been subtracted. The average mode for the opaques in the intrusion is 4.5 percent by weight, of which it is estimated that 3 percent is magnetite + titanomagnetite + ulvo-spinel, and the remaining 1.5 percent is ilmenite, the nonmagnetic fraction. As the average Ti content of the analyzed opaques in the intrusion is 6.5 percent, it has been assumed that their average Fe content is 63.5 percent.

(5) It is assumed that the ilmenite is an average one which contains 35 percent Fe and 30 percent Ti. The estimated values used are given in Table 11.

An examination of Tables 10 and 12 clearly reveals the location of many elements in the rocks; it also shows some of the problems faced in attempting mineral separations and analyses. The element content of the minerals has been totaled where it is believed that most of the element has been accounted for. Good balance between the element content of the minerals and that in the rock is shown by Co and Sc. These elements provide a good opportunity for the study of their geochemical behavior and partitioning, as they can be followed in all mineral locations in the intrusion. Where the element content of the minerals is considerably less than that of the rock, the element obviously entered an undetermined mineral phase, or formed an independent phase. For example, the discrepancy in Ba in rock W-824-60 indicates that much of the Ba presumably entered biotite. Furthermore, in all rocks, the determined minerals phases account for little of the total Cu and Zr, as these two elements formed independent minerals. In other cases, a reasonable balance between the element content in the minerals and that in the rock is apparent in the early-stage rocks, whereas a pronounced discrepancy occurs in the late-stage rocks. For example, Y behaves this way; the discrepancy indicated in the late-stage rocks results either from its entry into apatite or from the formation of a separate phosphate phase intergrown with apatite (*see* discussion of trace element behavior of Y).

The Cr, Ni, and V values indicate the difficulties encountered in mineral analysis. Contamination has been referred to already, and it is probably the explanation for the apparent excess of Cr and Ni in some mineral phases. The apparent error in Cr values results, it is believed, from impurities high

TABLE 12. ELEMENT PARTITIONING IN ROCKS

	Fe (%)	Mg (%)	Ca (%)	Mn (%)	Ti (%)	Ba (ppm)	Co (ppm)	Cr (ppm)	Cu (ppm)
COMPOSITION OF MINERALS (based on modal values)									
W-824-60	10.63	11.63	4.83	0.16	0.53	150	140	290	90
Σ Minerals	11.08	10.69	4.81	0.16	0.95		133	386	
Ol	4.56	5.35		0.079	0.012		92	4	4
Opx	4.0	3.78	0.34	0.056	0.059		28	92	7
Cpx	1.18	1.52	2.16	0.027	0.057		12	287	4
Pl	0.19	0.04	2.31	0.001	0.002	53		2	2
Mt–Il	0.20			0.001	0.007		1	1	1
Il	0.95				0.81				
W-U-60	7.35	5.13	8.18	0.12	0.61	150	52	430	95
Σ Minerals	8.19	5.6	8.03	0.12	0.61	125	55	704	
Opx*	4.83	3.1	0.7	0.063	0.054	17	32	122	5
Cpx	2.25	2.47	3.48	0.056	0.073		22	562	4
Pl	0.27	0.03	3.85	0.002	0.003	108		3	7
Mt–Il	0.32			0.003	0.029		1	17	1
Il	0.52				0.45				
W-N-60	7.67	3.06	7.01	0.12	0.74	210	42	16	120
Σ Minerals	8.05	2.0	7.15	0.12	0.51		43	18	
Pg	2.15	0.6	0.22	0.043	0.043		14	1	
Cpx	2.44	1.36	2.05	0.05	0.077		20	14	4
Pl	0.34	0.04	4.88	0.003	0.005	152		3	14
Mt–Il	3.12			0.029	0.38		9		7
W-E-60	14.2	2.29	5.73	0.19	2.14	320	75	tr	250
Σ Minerals	10.5	1.68	4.54	0.12	2.02		48	3	
Ol	1.43	0.07		0.016	0.013		5		2
Cpx	3.97	1.59	2.48	0.076	0.125		33	2	6
Pl	0.23	0.02	2.06	0.002	0.002	110		1	3
Mt–Il	3.12			0.023	0.38		10		26
Il	1.75				1.5				
COMPOSITION OF MINERALS (based on normative (N) and modal (M) values).									
W-824-60	10.63	11.63	4.83	0.16	0.53	150	140	290	90
Σ Minerals	10.53	10.68	4.86	0.16	0.94		132	331	
Ol	4.53	5.31	0.05	0.078	0.012		91	4	4
Opx	3.72	4.14	0.37	0.062	0.065		31	101	8
Cpx	0.9	1.18	1.68	0.021	0.045		9	223	3
Pl	0.23	0.05	2.76	0.001	0.001	64		2	2
Mt–Il	0.20			0.001	0.007		1	1	1
Il	0.95				0.81				
W-U-60	7.35	5.13	8.18	0.12	0.61	150	52	430	95
Σ Minerals	7.31	4.82	7.73	0.11	0.59	136	48	579	
Opx*	4.36	2.8	0.63	0.057	0.048	15	30	110	4
Cpx	1.8	1.98	2.78	0.045	0.058		17	449	3
Pl	0.31	0.04	4.32	0.002	0.003	121		3	8
Mt–Il	0.32			0.003	0.029		1	17	1
Il	0.52				0.45				

* Orthopyroxene including inverted pigeonite   tr Trace

Table 12. (Continued)

Ga (ppm)	La (ppm)	Ni (ppm)	Sc (ppm)	Sr (ppm)	V (ppm)	Y (ppm)	Zr (ppm)	Mode-Norm (Wt %)
COMPOSITION OF MINERALS (based on modal values)								
14	14	500	27	110	155	24	90	
		490	29	89	129	19		
		312	3		5	5		24
	3	118	12		40	6		23.6
	2	55	14	3	66	5		16.9
5	tr	2		86	2	3		25.2
tr	tr	3	tr		16	tr	1	0.3
								2.7
15	20	110	38	165	220	25	90	
		154	36	135	204	16		
	3	75	13		46	6		24.4
	4	76	23	4	112	6		28.1
9				131	4	4		43.0
tr	1	3	tr		42	tr	1	0.5
								1.5
20	21	40	30	250	235	26	115	
		52	29	200	424	19		
	2	14	6		26	3		8.6
	4	29	22	3	118	6		19.7
10				197	5	6		60.6
3	7	9	1		275	4	14	5
17	43	30	41	170	715	43	170	
		25	35	121	585			
	0.5	0.5	tr			1	4	2.8
	6.5	19.5	34	4	157	10		26.1
6				117	3	3		30.5
4	9	5	1		425	5	15	5
								5
COMPOSITION OF MINERALS (based on normative (N) and modal (M) values).								
14	14	500	27	110	155	24	90	
		486	27	104	116	15		
		309	3		5	5		23.8(N)
	4	129	13		44	6		25.9(N)
	2	43	11	2	51	4		13.14(N)
6		2		102				30.13(N)
tr	tr	3	tr		16	tr	1	0.3(M)
								2.7(M)
15	20	110	38	165	220	25	90	
		132	31	150	174	16		
	3	68	12		42	6		22.06(N)
	3	61	19	3	90	5		22.45(N)
10				147		5		48.28(N)
tr	1	3	tr		42	tr	1	0.5(M)
								1.5(M)

Table 13. Analyses of Rocks, Englewood Cliff and Union City Sections

Specimen	Height*	Fe (%)	Mg (%)	Ca (%)	Mn (%)	Ti (%)	Al (%)	B (ppm)	Ba (ppm)	Co (ppm)	Cr (ppm)	Cu (ppm)	Ga (ppm)	La (ppm)	Mo (ppm)	Nb (ppm)	Nd (ppm)	Ni (ppm)	Pb (ppm)	Sc (ppm)	Sn (ppm)	Sr (ppm)	V (ppm)	Y (ppm)	Zr (ppm)
											ENGLEWOOD CLIFF														
W-889LC-60(a)	1	7.89(c)	4.58(c)	7.39(c)	0.12(c)	0.73(c)	7.66(c)	tr	195	53	315	110	14	30	nd	nd	nd	95	11	37	1	175	235	29	120
W-865-60	30	8.13(c)	5.58(c)	7.95(c)	0.13(c)	0.70(c)	7.07(c)	tr	160	56	580	100	12	20	nd	nd	tr	135	10	38	2	150	230	27	115
W-824-60	70	10.63(c)	11.63(c)	4.83(c)	0.16(c)	0.53(c)	4.58(c)	nd	150	140	290	90	14	14	nd	nd	nd	500	18	27	1	110	155	24	90
W-804-60	90	7.76(c)	5.78(c)	7.74(c)	0.12(c)	0.64(c)	6.63(c)	tr	150	56	715	90	14	23	nd	nd	tr	135	13	39	1	160	225	22	120
W-U-60	215	7.35(c)	5.13(c)	8.18(c)	0.12(c)	0.61(c)	7.53(c)	tr	150	52	430	95	15	20	nd	nd	tr	110	22	38	1	165	220	25	90
W-R-60	365	7.62(c)	3.50(c)	7.59(c)	0.12(c)	0.69(c)	8.48(c)	tr	185	47	70	110	17	25	nd	nd	tr	65	8	33	2	200	215	27	120
W-N-60	560	7.67(c)	3.06(c)	7.01(c)	0.12(c)	0.74(c)	8.78(c)	24	210	42	16	120	20	21	nd	nd	tr	40	25	30	1	250	235	26	115
W-J-60	685	11.34(c)	2.38(c)	5.76(c)	0.15(c)	1.60(c)	6.32(c)	20	310	60	7	190	19	37	nd	nd	tr	30	17	39	1	185	490	38	180
W-E-60	805	14.20(c)	2.29(c)	5.73(c)	0.19(c)	2.14(c)	5.77(c)	28	320	75	tr	250	17	43	nd	tr	tr	30	36	41	1	170	715	43	170
W-D-60	830	13.57(c)	2.07(c)	4.48(c)	0.16(c)	2.10(c)	5.97(c)	40	375	62	tr	215	20	47	nd	tr	tr	30	19	37	2	175	525	47	200
W-FUC-60	900	7.5 (s)	4.2 (s)	7.2 (s)	0.12(s)	0.72(s)	7.65(a)		165	52	270	115		≈14	nd			110		36		150	260	25	105
W-F-60	790	10.49(c)	0.8 (c)	2.62(c)	0.15(c)	0.93(c)	5.96(c)	140	560	27	7	300	22	63	nd	tr	35	7	16	23	1	185	13	58	330
											UNION CITY														
W-WU3-61	990	10.92(c)	1.31(c)	4.04	0.16(c)	1.08(c)	5.98(c)		610	27	tr	300		63	10			7		27		180	11	58	290
16441	≈980	13.5 (s)	1.1 (s)	4.5 (s)	0.22(s)	1.4 (s)	5.9 (s)	57	550	40	tr	500	21	67	nd	tr	tr	8	19	31	1	190	15	58	280
W-WU10b-61	870	9.3 (s)	2.3 (s)	4.6 (s)	0.16(s)	1.5 (s)	7.1 (s)	70	270	41	tr	35	20	37	nd	nd	tr	35	16	34	2	220	360	40	215
W-WU17b-61		9.12(c)	4.55(c)	7.58(c)	0.16(c)	0.74(c)	7.27(c)		180	55	260	250		14	nd			85		39		170	250	27	125

(a) Average analysis
(c) Chemical determination
(s) Optical spectrograph determination

tr Trace
nd Not detected
* Height above base in feet

TABLE 14. ANALYSES OF OLIVINES, ORTHOPYROXENES, AND AUGITES, ENGLEWOOD CLIFF SECTION

Specimen	Height*	Fe (%)	Mg (%)	Ca (%)	Mn (%)	Ti (%)	Al (%)	Ba (ppm)	Co (ppm)	Cr (ppm)	Cu (ppm)	La (ppm)	Mo (ppm)	Ni (ppm)	Sc (ppm)	Sr (ppm)	V (ppm)	Y (ppm)	Zr (ppm)
OLIVINES																			
W-824-60	70	19(em)	22.3(em)		0.33(s)	≈ 0.05(s)	0.38	nd	385	18	17	< 14	10	1300	14	nd	20	20	nd
16441(uc)	≈ 980	51.3(o)	2.4(o)		0.56(s)	0.47(s)	0.57	40	175	≈ 8	60	25	25	14	21	nd	10	41	150
ORTHOPYROXENES																			
W-824-60	70	14.4(o)	16.0(o)		0.24(s)	0.25(s)	0.87	nd	120	390	30	≈ 14	nd	500	50	nd	170	25	nd
W-U-60	215	19.8(o)	12.7(o)		0.26(s)	0.22(s)	0.83	70	130	500	20	≈ 14	nd	310	55	nd	190	26	nd
AUGITES																			
W-865-60	30	6.1(s)	8.7(s)	12.4(s)	0.14(s)	0.24(s)	1.42	nd	68	3500	17	< 14	nd	280	70	20	295	19	nd
W-824-60	70	7.0(s)	9.0(s)	12.8(s)	0.16(s)	0.34(s)	1.02	nd	70	1700	24	≈ 14	nd	325	85	15	390	29	nd
W-804-60	90	7.8(s)	9.6(s)	13.2(s)	0.22(s)	0.3 (s)	1.41	nd	80	3200	17	15	nd	300	90	20	400	24	nd
W-U-60	215	8.0(s)	8.8(s)	12.4(s)	0.20(s)	0.26(s)	1.08	nd	78	2000	14	≈ 14	nd	270	83	15	400	22	nd
W-R-60	365	11.6(s)	8.3(s)	11.0(s)	0.26(s)	0.37(s)	1.13	nd	90	320	48	22	nd	185	105	15	530	34	nd
W-N-60	560	12.4(s)	6.9(s)	10.4(s)	0.25(s)	0.39(s)	1.04	nd	100	70	22	20	nd	150	110	15	600	30	nd
W-E-60	805	15.2(s)	6.1(s)	9.5(s)	0.29(s)	0.48(s)	1.10	nd	125	≈ 8	22	25	nd	75	130	15	600	38	nd

(em) Electron probe determination
(o) Universal stage determination
(s) Optical spectrograph determination
nd Not detected
(uc) Union City Section
* Height above base in feet

TABLE 15. ANALYSES OF PLAGIOCLASES AND APATITE, ENGLEWOOD CLIFF SECTION

Specimen	Height*	Fe (%)	Mg (%)	Ca (%)	Mn (%)	Ti (%)	Ba (ppm)	Co (ppm)	Cr (ppm)	Cu (ppm)	Ga (ppm)	La (ppm)	Mo (ppm)	Ni (ppm)	Sc (ppm)	Sr (ppm)	V (ppm)	Y (ppm)	Zr (ppm)
PLAGIOCLASES																			
W-865-60	30	0.74(s)	0.14(s)	8.45(g)	0.004(s)	0.007 (s)	275	nd	7	16	15	nd	≈ 5	5	< 9	325	≈ 9	≈ 10	< 50
W-824-60	70	0.77(s)	0.17(s)	9.15(g)	0.005(s)	0.007 (s)	212	nd	7	8	19	nd	≈ 5	7	< 9	340	≈ 9	≈ 10	< 50
W-804-60	90	0.82(s)	0.09(s)	8.80(g)	0.005(s)	0.007 (s)	255	nd	9	15	18	nd	≈ 5	< 5	< 9	345	≈ 9	≈ 10	< 50
W-U-60	215	0.64(s)	0.08(s)	8.95(g)	0.004(s)	0.007 (s)	250	nd	7	17	20	nd	≈ 5	< 5	< 9	305	≈ 9	≈ 10	< 50
W-R-60	365	0.57(s)	0.07(s)	8.8 (g)	0.006(s)	0.008 (s)	245	nd	6	16	16	nd	≈ 5	< 5	< 9	305	≈ 9	≈ 10	< 50
W-N-60	560	0.56(s)	0.07(s)	8.05(g)	0.005(s)	0.0075(s)	250	nd	6	23	17	nd	≈ 5	< 5	< 9	325	≈ 9	≈ 10	< 50
W-E-60	805	0.76(s)	0.05(s)	6.75(g)	0.005(s)	0.008 (s)	360	nd	5	10	19	nd	≈ 5	< 5	< 9	385	≈ 9	≈ 10	< 50
16441(uc)	≈ 980	0.58(s)	0.02(s)	5.3 (g)	0.007(s)	0.0045(s)	750	nd	5	14	24	nd	≈ 5	< 5	< 9	395	≈ 9	≈ 10	< 50
APATITE																			
16441(uc)	≈ 980	0.3 (s)			0.05 (s)		48	9	16	28		3800	15	12	13	275	17	3600	200

(g) R.I. determination of fused plagioclase
(s) Optical spectrograph determination
nd Not detected
(uc) Union City Section
* Height above base in feet

TABLE 16. ANALYSES OF OPAQUE IRON MINERALS, ENGLEWOOD CLIFF SECTION

Specimen	Height*	Mn (%)	Ti (%)	Co (ppm)	Cr (ppm)	Cu (ppm)	Ga (ppm)	La (ppm)	Mo (ppm)	Ni (ppm)	Sc (ppm)	V (ppm)	Y (ppm)	Zr (ppm)
W-865-60	30	0.45(s)	6.0(s)	160	230	250	50	120	55	400	23	3800	75	300
W-824-60	70	0.38(s)	2.2(s)	250	225	425	45	100	50	1100	13	5300	50	380
W-804-60	90	0.30(s)	4.5(s)	200	1750	155	65	110	50	525	18	7000	55	260
W-U-60	215	0.59(s)	5.7(s)	225	3400	250	60	120	65	600	18	8500	55	270
W-R-60	365	0.56(s)	6.6(s)	165	50	130	60	160	55	375	20	5600	80	290
W-N-60	560	0.57(s)	7.5(s)	180	nd	135	55	140	60	175	18	5500	70	270
W-E-60	805	0.46(s)	7.6(s)	200	nd	120	80	170	60	100	18	9500	90	300
16441(uc)	≈ 980	0.62(s)	6.4(s)	125	nd	525	50	90	65	20	23	100	70	280

(s) Optical spectrograph determination
nd Not detected
(uc) Union City Section
* Height above base in feet

in Cr which contaminate the clinopyroxenes separated for analyses. This is probably also the cause of the Ni discrepancies indicated. Greenland and Lovering (1966) refer to NiS and CoS blebs that they detected by electron microprobe in the opaque iron minerals of the Great Lake Dolerite Sheet, and mineragraphic work in this study has indicated the possibility that finely divided opaque mineral inclusions are disseminated in some of the silicate mineral phases.

The discrepancy shown by V in rock W-N-60 could indicate that the modal value for the opaque iron mineral is too high, or that part of the opaque mineral is ilmenite, for the sum of Fe in the minerals is too high and Ti too low relative to the element values in the rock.

Element Content of Rocks and Minerals

To facilitate the discussion in the next section of the geochemical behavior of elements during the crystallization of tholeiitic magma, the major, minor, and trace element contents of the rocks and minerals are presented in order of progressive fractionation stages in Table 13 to 16, and the distribution plots of the major and minor elements are shown in Figures 13 to 18.

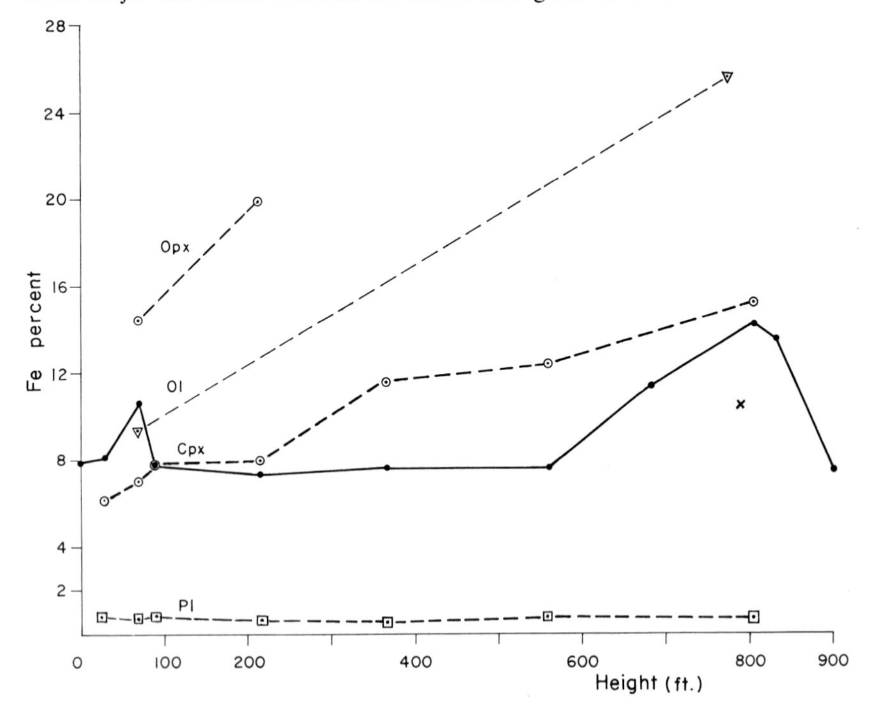

Figure 13. Distribution trends of Fe, Englewood Cliff Section. Iron in Ol shown at half scale.

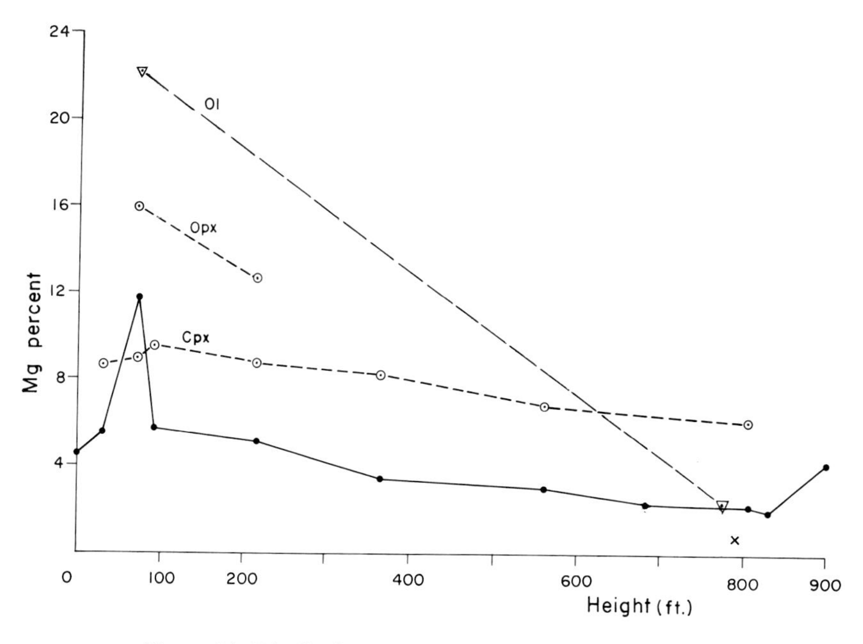

Figure 14. Distribution trends of Mg, Englewood Cliff Section.

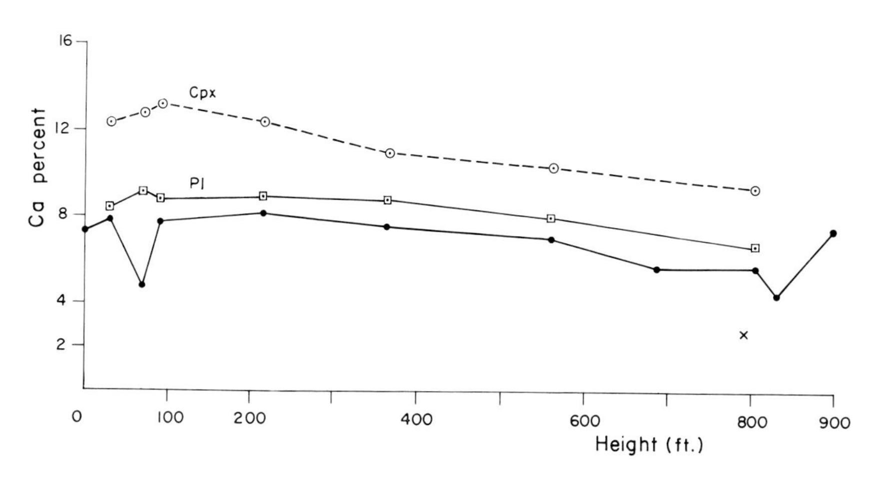

Figure 15. Distribution trends of Ca, Englewood Cliff Section.

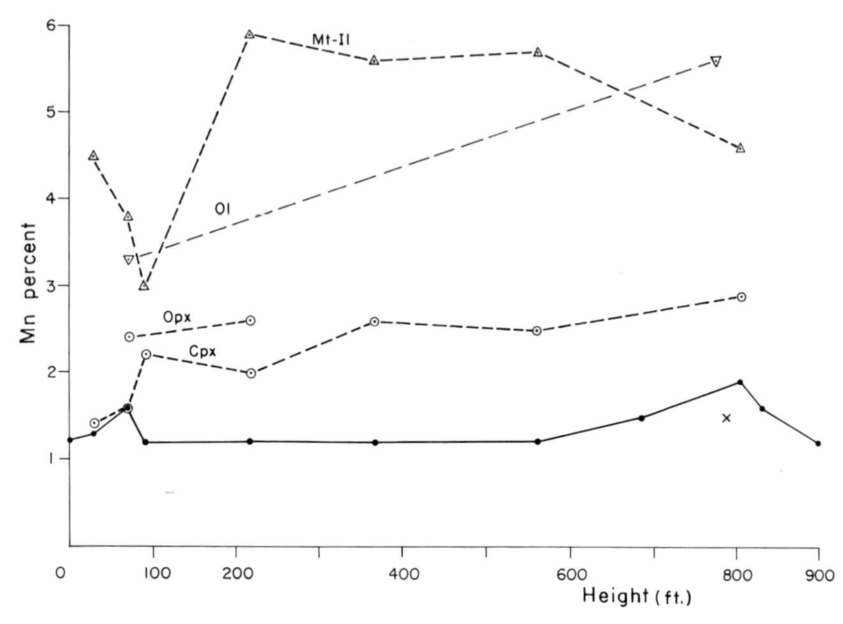

Figure 16. Distribution trends of Mn, Englewood Cliff Section.

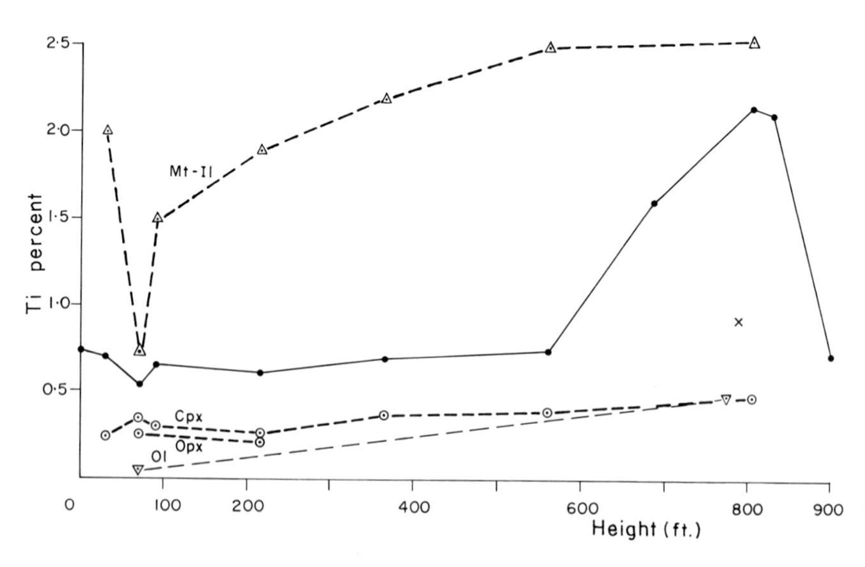

Figure 17. Distribution trends of Ti, Englewood Cliff Section. Titanium in Mt-Il shown at one-third scale.

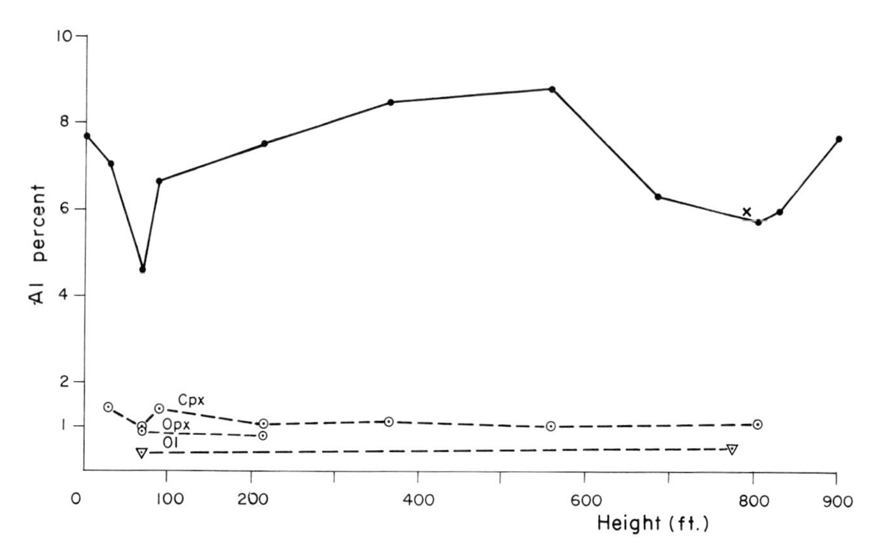

Figure 18. Distribution trends of Al, Englewood Cliff Section.

TRACE ELEMENT CHEMISTRY

The Palisades Sill provides an excellent opportunity for the study of trace element partitioning and behavior in a fractionated tholeiitic intrusion because, in its distinct mineralogical layers, the element distributions fluctuate sharply, and because coherence relationships can be followed over a wide compositional range in each mineral series.

The procedure followed in the presentation of element distribution plots has been detailed. The coherence relationships are plotted using logarithmic scales in Figures 21, 25, 26, 29, 30, 32, and 35. These show the variation in the trace element content (ppm) of rocks and minerals with the number of lattice sites available during fractionation; the sites are those normally entered by the major element that the trace element follows in its distribution. The number of available lattice sites is taken as proportional to mole of the major element, that is, to the weight of the major element in 100 gms of rock divided by its atomic weight. In these diagrams the coherence relationships in the whole rocks are shown by a full line, and those in the minerals by a broken line. All coherence diagrams use the same symbols; numbering and lettering of points on curves refer to specimen numbers. For example, 824, E, and 104 refer to specimens W-824-60, W-E-60, and 14-104 respectively. Where points are unmarked in the diagrams, they may be identified by comparing them with corresponding labeled points.

Because the atomic weights of the elements within a particular coherence group are close, the slopes of the coherence curves serve as a reasonable guide to the ease and order of entry of the trace cations into major cation lattice sites. For example, the competition by Cr, V, and Sc (Atomic weights 45 to 52) for sites in crystal lattices normally occupied by Fe, Mn, and Ti can be compared on this basis. Thus it can be seen from the coherence curves that Cr (low slope) entered lattice sites of minerals with fractionation much more rapidly than V, and in turn, V entered somewhat more rapidly than Sc (high slope).

Geochemical 'rules' as originally defined by Goldschmidt (1937, 1944) and later added to, or modified by, other workers (Ringwood, 1955a and b; Ahrens, 1963) have served as a useful guide to geochemists for explaining the behavior of some trace elements. To date, most attempts by geochemists to explain element behavior have been based on considerations of ionic or physical properties (ionic size,[3] electronegativity, charge, and so on) of the ion in a single solid phase, and on bond-forming characteristics, particularly with oxygen. This has obviously been too limited an approach because in some cases predictions based on one, or a combination, of those properties considered so far have not been in accordance with observed behavior.

It appears that the main limit applied in using the 'rules' to explain geochemical behavior is that allowance is not always made for the variations that occur in the ionic properties of an ion between phases in a magma, and thus for the effects that these have on the thermodynamics of the reactions that take place. In the present study it has been observed, for example, that Co mainly follows Fe^{2+} in clinopyroxene and orthopyroxene, and Mg in olivine, and thus that the bonding characteristics of these ions vary somewhat in the different lattice structures. This is apparently due to the difference in the configuration of the surrounding ions in each lattice. It is not possible to determine the ionic properties of elements in the liquid and solid phases involved in such reactions and discuss them in detail, but the study does reveal the distribution of elements between mineral phases with fractionation, from which the factors involved in element partitioning have been evaluated.

In the magmatic environment the ions of both the major and trace elements appear to tend toward the most stable and ordered arrangement possible during the solidification of the intrusion, and it is the major elemental composition of the magma which largely determines the types of minerals and the order in which they form. In the vicinity of a forming crystal, ionic activity would be expected to vary according to the liquid composition and temperature, factors which would influence the type and number of ligands forming. A free ion in this situation would be incorporated if the thermodynamics and its ionic properties favored its entry into a lattice in competition with other available ions. The ions of an element appear to show preference for particular sites within a mineral lattice. The extent of entry will also depend on the concentration of the ions of the element in association with other ion concentrations in the liquid, particularly those of the major elements with which it associates, and on the degree of its preferred entry over the other competing ions.

From the element distributions observed in the Palisades it can be said that, *in general terms*, the cation of a trace element tends to enter the mineral

---

[3]The ionic radii (in Ångstrom units) quoted throughout are from Pauling (1960, p. 514-518). The ionic radius is given in parenthesis immediately after the cation to which it refers.

lattice site normally occupied by a major cation whose chemical properties are similar; for example, where Sr^{2+} enters a Ca^{2+} site. Where differences in physical properties, such as ion size, militate against this, the trace cation tends toward a physically possible site normally occupied by a major cation whose chemical properties are closest; for example, where Sr^{2+} enters a K^+ site. Some trace elements are only accommodated with difficulty in the lattice sites of the major mineral-forming elements, and these tend to behave independently; for example, Zr and Cu. Thus, where trace elements have no similar major elements with which they can associate, they tend to concentrate in the magma until they can form a separate mineral phase. Their behavior then switches from that of a trace to that of a major mineral-forming element, with which other suitably matched trace elements may then associate.

Apparently gravity, like pressure, has little influence in the liquid phase on element partition equilibrium. Magnesium, for example, was selectively removed in preference to Fe during progressive crystallization, and the heavier Fe ions concentrated toward the top of the intrusion, whereas the lighter Mg ions concentrated toward the base. The controls on element behavior have an important bearing on the differentiation of the intrusion. However, gravity is important in the differentiation of solid phases, and if convection occurs, it also is in the liquid phase. Factors in the differentiation process are discussed in the section on petrogenesis.

Terms used to convey ideas consistent with the concepts presented in this paper have meant abandoning such geochemical terms as diadochy, captured, and so on, because of their implication on element behavior and their association with the classical "rules." In this paper where an element is described as "entering lattice sites," it implies that the element occupies the sites that might normally be occupied by a major cation in the lattice structure. That is, the trace cation successfully competes with the major cation for the available site. For example "Co enters Mg sites in olivine" means that Co successfully competed with Mg and occupied those sites in the olivine structure which are normally occupied by Mg because its chemical properties in the magnetic system favored the reaction.

The term "coherence" is retained, as it is a useful term to convey the idea of a close distribution relationship existing between elements of similar behavior, and, as a trace element appears to show site preference in a lattice and enters sites on a proportional basis relative to its concentration and to that of the major element with which it associates, it is valid to say that the trace element follows the major element in its distribution.

Conclusions reached on element behavior from whole-rock studies alone can be misleading. For an element coherence relationship to be valid, it must be substantiated in at least one mineral series of the fractionated rock series.

Thus, it is necessary first to study element partitioning and identify relationships in all the mineral series of the rock, and then, from the results, interpret the whole-rock element relationships, after making the appropriate modal corrections for mineral content.

The following discussion of element distribution and partitioning in the Palisades Sill reinforces the general concepts outlined above on element behavior. The results of the trace element studies are presented according to element behavior with fractionation in the intrusion. This behavior, it has been found, corresponds closely to predictable chemical behavior, so that the arrangement of elements for discussion in most cases conforms with a natural grouping of the elements consistent with the Periodic classification.

Strontium and Barium

Strontium. Strontium in the original magma was 175 ppm, and the average composition of the intrusion is about 190 ppm. The distribution of Sr in the whole rocks and minerals is shown in Figure 19, and the values are presented in Tables 13 to 15. The minimum concentration of 110 ppm occurs in the Mg-olivine layer, and during the early fractionation stages, Sr content built up slightly in the magma. Strontium was mainly removed during the middle stages, where content in the whole rocks reached the maximum of 250 ppm in the late pigeonite dolerite. The magma was consistently impoverished in Sr thereafter with fractionation.

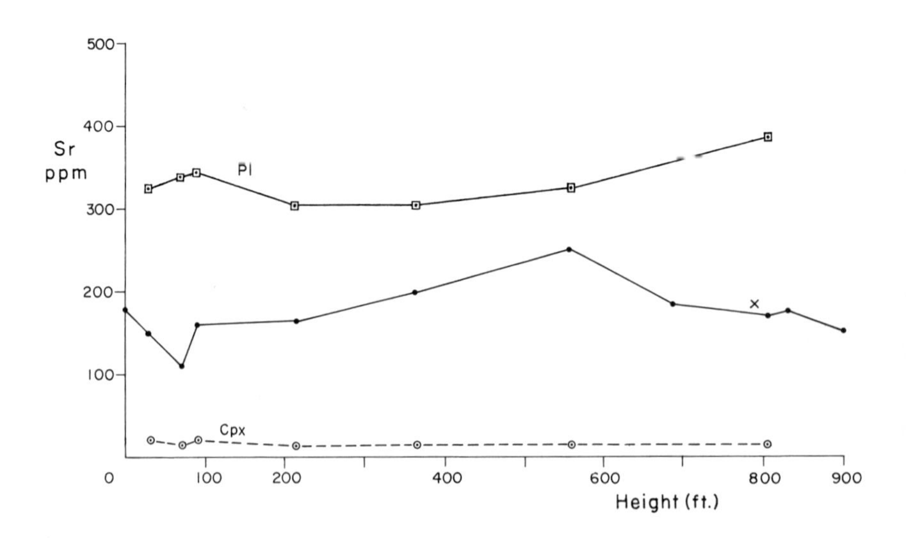

Figure 19. Distribution trends of Sr, Englewood Cliff Section.

Most of the Sr entered plagioclase; apatite has similar Sr values to plagioclase, but is a minor accessory. Only trace amounts of Sr are present in the clinopyroxene, and it is absent from olivine, orthopyroxene, and the opaque iron minerals.

The control of Sr distribution in plagioclase is not conclusively established. Strontium concentration in the whole rocks varies consistently with the modal content of plagioclase to the pigeonite dolerite stage. Figure 21 shows the relationship of log Sr (ppm) to log plagioclase (volume percent). No relationship is apparent between Sr and Ca alone in plagioclase. Strontium content shows a steady increase with fractionation in the middle and late fractionation stages, whereas Ca content progressively decreases. This increase may be due to the increasing number of K sites available (that is, sites of the type normally occupied by K),[4] as the plagioclase became richer in alkali. Possibly, also, the plagioclase separated for analysis from the products of late fractionation stages contains some alkali feldspar contamination. However, it appears that some Sr entered Ca sites (that is, sites of the type normally occupied by Ca),[4] as the most Ca-rich plagioclases in the early fractionation stages, and, in particular, that in the Mg-olivine layer, contain more Sr than the plagioclases in the middle stages.

Though only trace amounts of Sr occur in the clinopyroxene, its distribution in early fractionation stages suggests that it also entered Ca sites (that is, those normally occupied by Ca),[4] The Sr content in the clinopyroxene of the Mg-olivine layer is somewhat less than that in the clinopyroxenes immediately above and below the layer, which are probably richer in Ca.

In the whole rocks Sr follows (K + Ca) until the late hypersthene dolerite stage of fractionation (Fig. 21). Thereafter, it appears to follow only K until the late pigeonite dolerite stage. These coherence relationships cannot be followed in the late fractionation stages, and probably lapsed because there was insufficient Sr in the magma to maintain them. In late stages, apatite, and hornblende are more abundant and probably provided additional sites for Sr, and thus further complicate the problem of elucidating Sr behavior.

Barium. The original magma contained 195 ppm Ba, whereas the average composition of the intrusion is 21·5 ppm. Barium was concentrated in the magma during the early and middle fractionation stages, and was rapidly removed in the late stages, though it appears to maintain coherence relationships

---

[4]For brevity, the full explanation of element entry into a crystal lattice as given here in parenthesis is not repeated in the discussion of the elements that follow. The site into which the trace enters is described by the major or minor element that normally occupies it.

throughout and follows K distribution closely (Fig. 21). The late-stage alkali-enriched rocks, such as the granophyric dolerite, contain up to 600 ppm Ba. The formation of the Mg-olivine layer had little effect on Ba distribution in the magma. The distribution trends in the whole rocks and minerals are shown in Figure 20 and values are given in Tables 13 to 15.

Most of the Ba occurs in feldspar and biotite, though small amounts are in the Fe-rich olivine and Fe-rich hypersthene, and in the apatite. It was not detected in the clinopyroxenes and opaque iron minerals nor in the Mg-rich olivine and Mg-rich orthopyroxene. Barium distribution in the whole rocks during middle fractionation stages is controlled by the modal distribution of plagioclase. In early, and, in particular, in late stages, biotite, apatite, hornblende together with alkali feldspar apparently also provided some of the lattice sites used by Ba, so that the whole-rock relationship between Ba and modal plagioclase is not maintained throughout the intrusion.

The sharp increase in the Ba content of plagioclase in the late fractionation stages probably resulted from plagioclase contamination by alkali feldspar and from the possible increase in the K content of the plagioclase with fractionation. Contamination would possibly occur, as mentioned, because the alkali feldspar is intimately associated with plagioclase in the micropegmatite areas of crystallization, particularly in the late-stage dolerites.

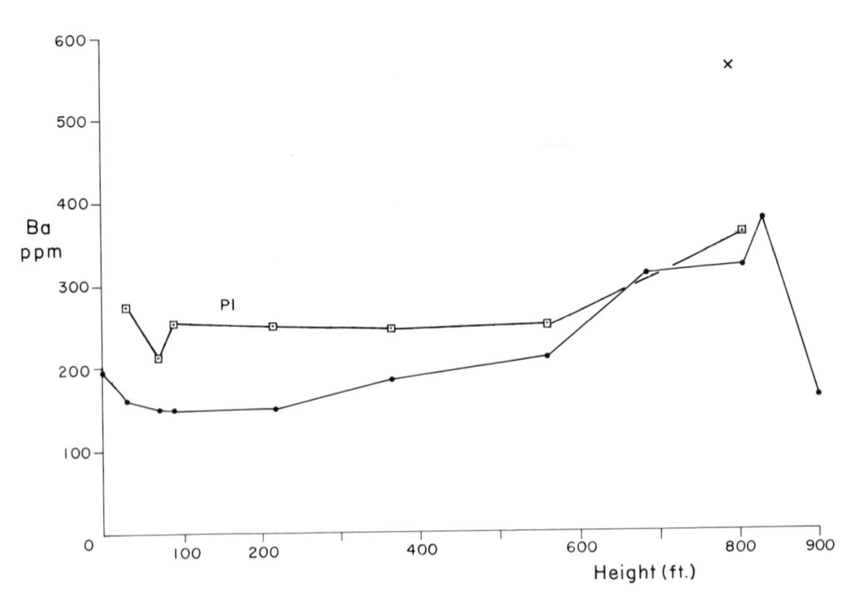

Figure 20. Distribution trends of Ba, Englewood Cliff Section.

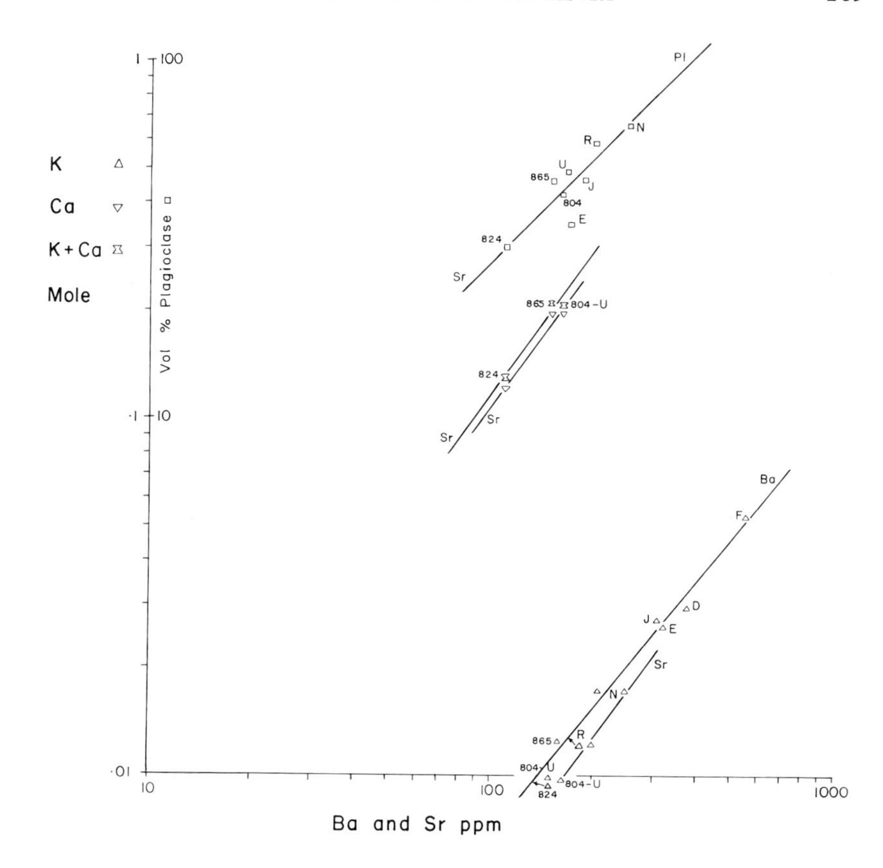

Figure 21. Coherence curves of Ba and Sr, and the relationship between Sr concentration and the modal distribution of plagioclase.

The Ba in the Mg-olivine layer apparently entered biotite, because its concentration in the layer is similar to that in the dolerites above and below, but the Ba content in the plagioclase is distinctly lower, and Ba is absent from all other minerals in the layer.

Geochemical behavior. The ionic radius of Sr^{2+} (1.13) is between K^+ (1.33) and Ca^{2+} (0.99) in size, whereas Ba^{2+} (1.35) is very close to K^+. Chemically Sr and Ba might be expected to follow mainly Ca, but neither conveniently enter Ca sites in minerals, except possibly those in apatite whose Ca sites are relatively large. Ion size apparently has an important influence on their geochemical behavior, and, even though they appear to have better bonding characteristics than Ca in many minerals, they discriminate between sites according to mineral type and preferably enter K sites. However, they are also influenced to some extent by Ca distribution. Most of the Ba and Sr entered plagioclase and potash feldspar and these largely control their distribution in the whole rocks. Barium also freely entered biotite.

It is difficult to completely resolve the behavior of Sr and Ba at present as K determinations are available only for the whole rocks, and all the coherence relationships indicated cannot be verified in the mineral series. Barium provides the better opportunity for the study of these relationships because most of it remained in the magma until late fractionation stages whereas Sr was largely removed in the middle stages.

Because the atomic weight of Sr is about .66 that of Ba, the slopes of the coherence curves imply that from early to middle fractionation stages at least, Ba obtained more favorable entry into K lattice sites than Sr, and that Sr entry in the early fractionation stages was about equally divided between Ca and K sites. However, the over-all distribution of Sr suggests that it preferred the K sites in minerals, including plagioclase in late stages because in this mineral series Sr values progressively rise and Ca fall with fractionation. The charge balance problem resulting from Sr^{2+} or Ba^{2+} entry into K^{+} sites may have been overcome by Al^{3+} taking the place of Sr^{4+} in the lattice.

The somewhat more favorable entry of Ba than Sr into K sites was probably aided by the similarity in the size of the Ba and K ions and by the better bonding of Ba than Sr indicated in lattices. The removal of Sr before Ba from the magma during fractionation may result from Sr entering K and Ca sites in early- and middle-tage plagioclases whereas Ca sites in the plagioclase lattice did not provide convenient entry for Ba. Barium apparently entered K sites in both feldspar and biotite and these progressively increased in numbers toward late fractionation stages. Concentration of Ba toward these late stages would also result if Ba has weaker bonding characteristics in mineral lattices than K.

Strontium apparently entered Ca sites in clinopyroxene. Barium appears also to have entered Ca sites to a limited extent in the orthorhombic pyroxene crystal structure, whereas, it could not enter the monoclinic structure. It was not detected in clinopyroxene, but it did enter the Fe-rich olivine and Fe-rich hypersthene. Neither olivine nor orthopyroxene would be expected to have much K, but the Fe-rich varieties could be expected to contain more Ca than the Mg-rich varieties. The Ba values (Table 14) imply that the Palisades-type orthopyroxene contains nearly twice as much Ca as the fayalite, and that the Bushveld-type orthopyroxene and the hyalosiderite have low Ca contents.

Strontium entered apatite much more abundantly than Ba, and both probably occupy Ca sites. Entry into apatite would have been assisted by the large lattice site of the Ca cation, for Ca is present in 9-fold co-ordination

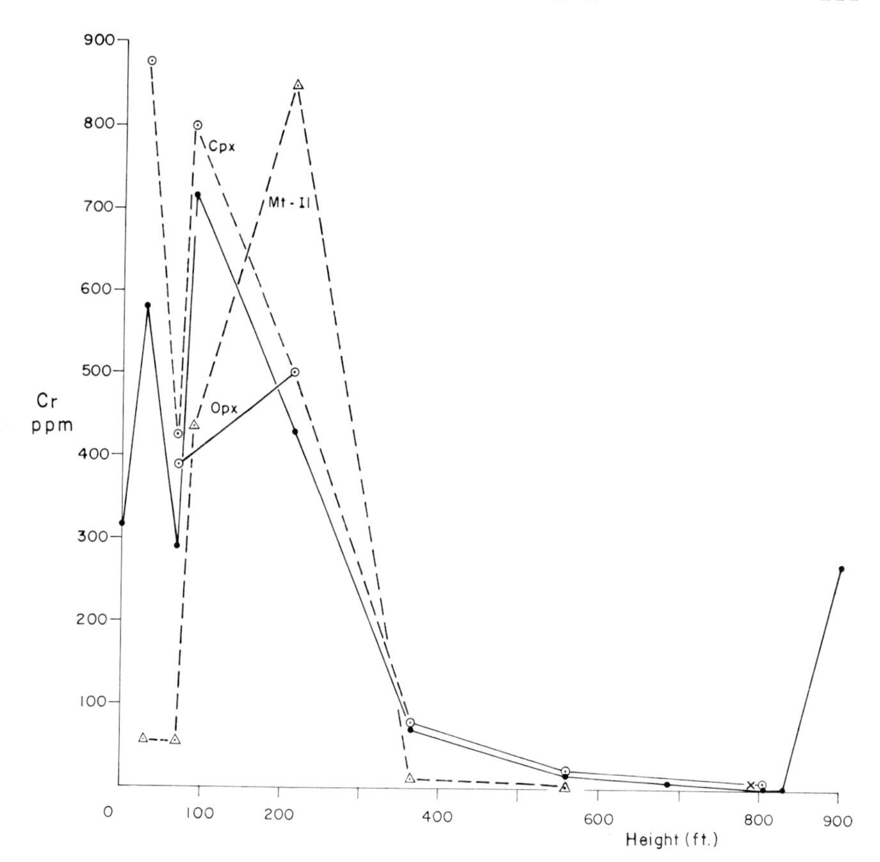

Figure 22. Distribution trends of Cr. Englewood Cliff Section. Chromium in Cpx and Mt-Il shown at one-quarter scale.

with oxygen. Again the more favorable entry of Sr would be expected on size considerations.

Chromium, Vanadium, and Scandium

Chromium. Chromium was rapidly removed from the magma in the early fractionation stages. The distribution trends of Cr in rocks and minerals are shown in Figure 22, and element values are given in Tables 13 to 16. The original Palisades magma contained 315 ppm Cr, whereas the average Cr content of the intrusion is 190 ppm.

Chromium freely entered trivalent lattice sites, and its distribution in the whole rocks with fractionation closely parallels that in the clinopyroxene mineral series. However, because of early depletion of the magma in Cr, the relationship between Cr in the whole rocks and the modal content of clinopyroxene is barely maintained, even in the early fractionation stages. In later stages it lapsed completely, and it is impossible to follow Cr coherence with major cations in crystal lattices throughout fractionation. The effect of variations in the modal content of the minerals on Cr distribution in the whole rocks is demonstrated by the sharp decline in Cr to 290 ppm in the Mg-olivine layer compared to the maximum of 715 ppm in the bronzite dolerite immediately above. This decline may be explained by a marked decrease in the modal content of clinopyroxene in the hyalosiderite dolerite, whose other mineral phases, except the opaque iron minerals, were unfavorable for Cr entry. Moreover, the clinopyroxene of the layer (with 1700 ppm Cr) apparently contained less suitable Fe sites for entry than did the clinopyroxenes (with more than 3000 ppm Cr) in the dolerites immediately above and below.

The rapid depletion of the magma in Cr took place during the hypersthene dolerite and early pigeonite dolerite stages; most of the Cr was removed by the late pigeonite dolerite stage, and only traces are present in the ferrodolerites and later differentiates. Reappearance of Cr in late-stage fractionation products, as reported by Wager and Mitchell (1951) in the Skaergaard and by Greenland and Lovering (1966) in the Great Lake Sheet does not occur in the Palisades intrusion.

Chromium entered clinopyroxene abundantly, and had nearly as favorable an entry into the opaque iron minerals, where it probably mainly entered magnetite. Much less entered orthopyroxene, and only traces are found in olivine. In the opaque iron minerals it reached peak concentration in the hypersthene dolerite stage, where the opaques are modally the lowest in the intrusion. However, as mentioned, the opaques selected for analysis are not necessarily completely representative of each fractionation stage. The possible coherence relationship with Al in early clinopyroxenes and early rocks, referred to below, and the problem of contamination in the clinopyroxenes separated for analysis, further complicate the discussion of Cr behavior.

Chromium distribution in the minerals indicates that it has strong coherence with Fe, Ti, and Mn, and the low slope of curves relating Cr distribution with these elements (Fig. 25) further supports this by indicating the rapid entry of Cr with fractionation. Distribution in whole rocks reveal coherence with Fe and Ti only, as apparently the variation in Mn concentration is insufficient for the relationship with Cr to be revealed.

The original magma contained 460 ppm Cr_2O_3, though no Cr-spinel has been seen in the intrusion. This value is somewhat higher than that (250 ppm) stipulated for the separation of Cr-spinel in the Skaergaard by Wager and Mitchell (1951), but, of course, the environment of crystallization is very different in the Palisades, and physico-chemical conditions are too poorly known to interpret the significance of Cr concentration in terms of Cr-spinel crystallization.

Vanadium. The concentration of V built up in the magma during the early and middle fractionation stages and depletion did not occur until after the ferrodolerite stage. The original magma contained 235 ppm V, and the average composition of the intrusion is 315 ppm. Distribution of V in the rocks and minerals is shown in Figure 23, and values are given in Tables 13 to 16. The maximum V concentration of 715 ppm was reached in the ferrodolerites, and the minimum, excluding the late-stage granophyric rocks, is 155 ppm in the Mg-olivine layer. Both Sc and Cr show similar sharply reduced concentrations in the layer, a fact supporting common control in their behavior.

Vanadium in the mineral phases follows Fe, Ti, and Mn with fractionation. The clinopyroxene series demonstrates this coherence clearly (Fig. 26). the slopes of the curves showing these relationships are steeper than the

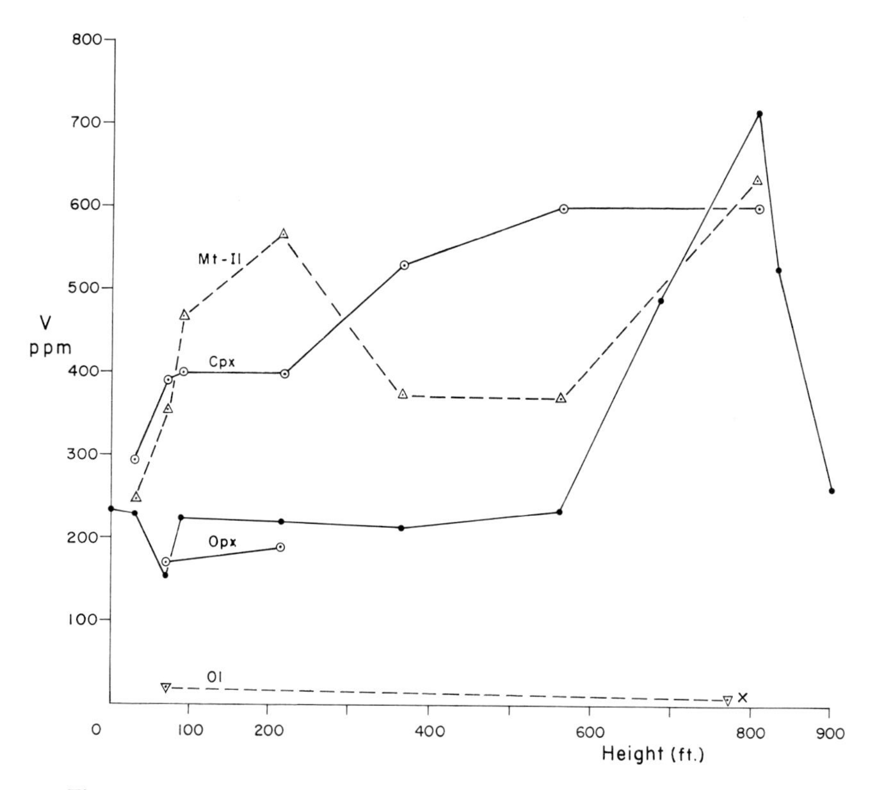

Figure 23. Distribution trends of V, Englewood Cliff Section. Vanadium in Mt-Il shown at one-fifteenth scale.

corresponding ones for Cr and indicate that V with fractionation did not enter mineral lattice sites as freely as Cr. Similar relationships are shown by V to Fe, Ti, and Mn in orthopyroxene, which it entered abundantly, but not in olivine, for the fayalite granophyre, whose olivine was analyzed, formed after the magma was depeted in V. Vanadium distribution in the opaque iron minerals follows Ti; it would be expected that Fe analyses of the opaques would reveal a strong coherence between V and Fe. The abundance of V in the opaques indicates that those separated are largely magnetite, which V freely entered.

Vanadium distribution in the whole rocks is governed by the modal distribution of clinopyroxene and the opaque iron minerals. Vanadium coherence with Ti and Fe was maintained until the late fractionation stages, but in very late stages the magma was impoverished in V and the relationship lapsed. In the early fractionation stages the strongest coherence appears to be with Ti; during the middle and late stages very good coherence is shown with (Ti + total Fe). From the whole rock coherence relationships no dis-

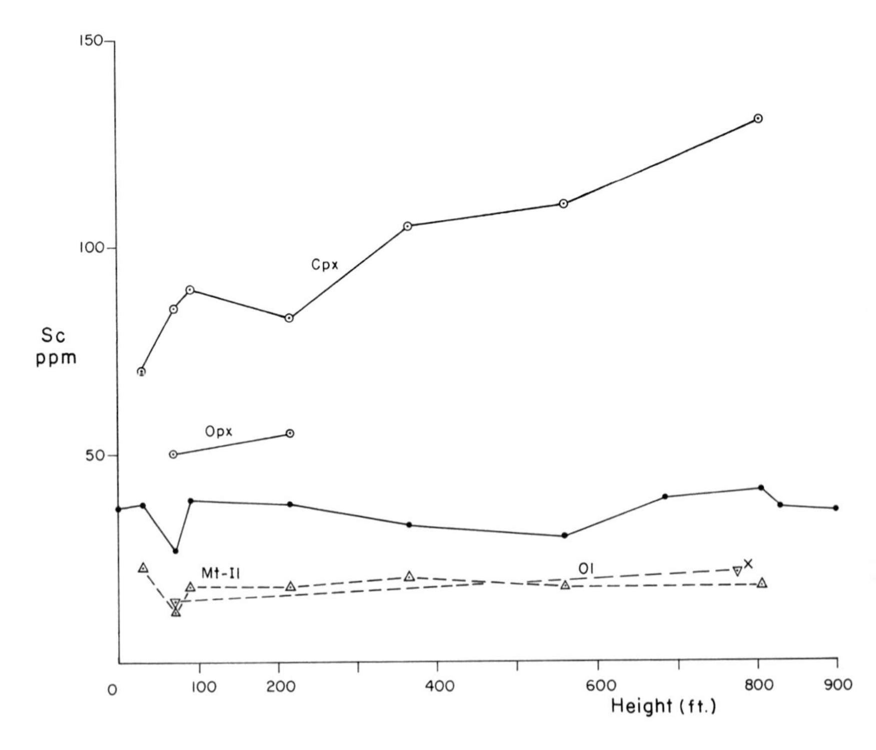

Figure 24. Distribution trends of Sc, Englewood Cliff Section.

tinction is apparent in the preference of V for Fe^{3+} or Fe^{2+} sites. The fairly constant Mn content of the whole rocks, except in the late-stage rocks, apparently masks the relationship with Mn revealed in the mineral series.

Scandium. Though the range in concentration of Sc in the intrusion is only 27 to 41 ppm, the distribution trends in the whole rocks and minerals are similar to those of V, except for the opaque iron minerals (Fig. 24). Also like V, concentration of Sc tended to build up in the magma during early and middle fractionation stages, and Sc was removed from the magma during the late stages. However, the magma was impoverished in Sc at a somewhat later stage than for V, and the granophyric dolerite contains 23 ppm, whereas the V content by this and the fayalite granophyre stage was reduced to about 10 ppm. The initial magma contained 37 ppm Sc, and the average for the intrusion is about the same. The Sc content of the rocks and minerals are given in Tables 13 to 16.

Scandium, like Cr, shows preferred entry into clinopyroxene, which it entered in increasing amounts throughout fractionation. Moreover, the low Sc content of the Mg-olivine layer relative to that in the dolerites above and below results from the low modal content of clinopyroxene in the layer. In contrast to V, scandium does not appear to have entered the opaque iron minerals very abundantly, but this may be because, as mentioned, the method of separation used to obtain the opaques excluded much of the ilmenite, and Sc shows good coherence with Ti. Some coherence with Fe in the opaques would also be expected.

Scandium entered mainly Mn and Ti lattice sites, and also those normally occupied by Fe, including Fe^{2+}. This is shown by the coherence between Sc and Mn, and Sc and Ti, in the clinopyroxene, orthopyroxene, and olivine mineral series. A relationship between Sc and (Ti + total Fe) is also indicated. The distribution trends of Sc and Fe in clinopyroxene are similar (Figs. 13 and 24). Slopes of the coherence curves suggest that Sc entered Mn and Ti sites in clinopyroxene with equal ease, and that entry into Fe sites was somewhat less favored. This is a similar pattern to that shown by V, except that the Fe sites were those most favored by V.

Because the magma was depleted in Cr and V in the very late fractionation stages, their coherence relationships in olivine could not be studied. The somewhat later depletion of the magma in Sc enables its behavior in olivine to be followed; scandium entered Mn and Fe sites, but these sites were less acceptable than those in the pyroxenes.

The coherence relationships of Sc in the whole rocks is largely controlled by the modal distribution of clinopyroxene, and, to a much lesser extent, by the opaque iron minerals. Scandium coherence is with Ti and Fe, though entry into Fe positions was obtained with much greater difficulty than with

V. Where probably V preferred Fe^{3+} sites, Sc possibly showed a slight preference for Fe^{2+} sites. Main Sc coherence in early fractionation stages is with Ti, but during the middle and late stages it is with (Ti + total Fe). In the very late stages some magma impoverishment in Sc developed, and the relationship between Sc and (Ti + total Fe) lapsed for this reason. The variation in concentration of Sc and Mn with fractionation in the whole rocks does not reveal the relationship between them.

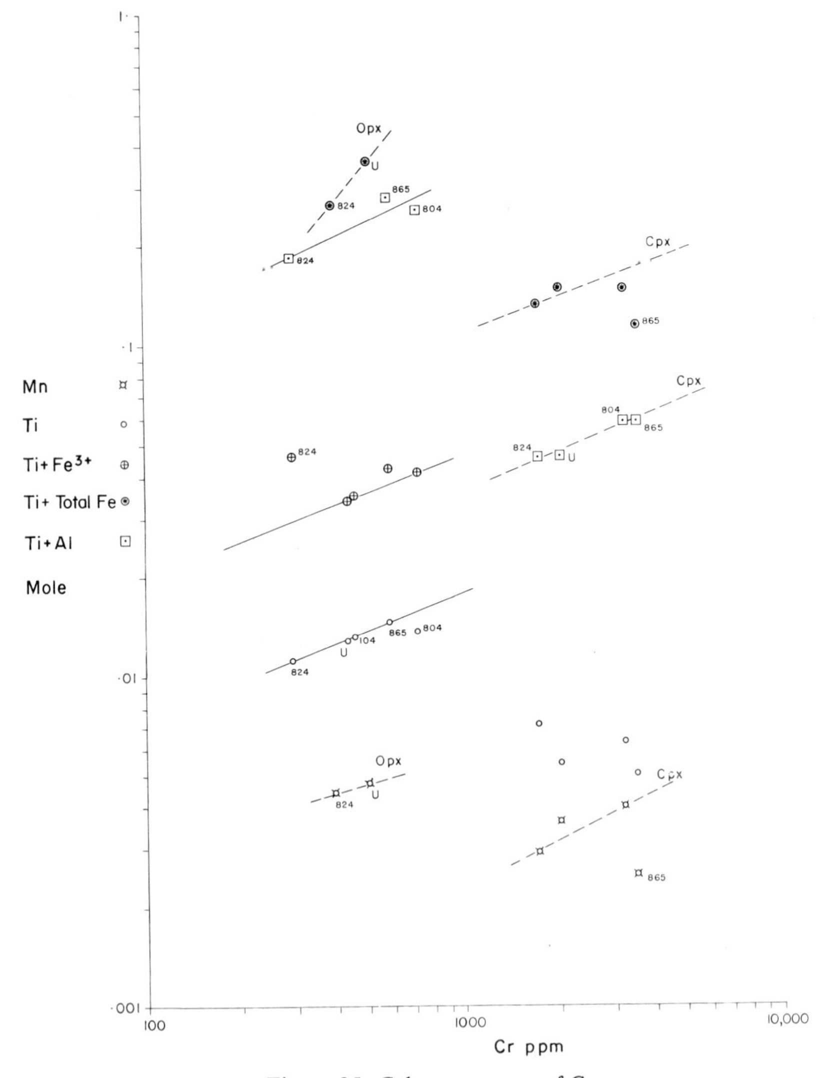

Figure 25. Coherence curves of Cr.

Geochemical Behavior. Chromium, V, and Sc have similar chemical properties and show similar geochemical behavior. Chromium, however, is complementary to V and Sc in its distribution trends. The favorable mineral lattice sites for Cr^{3+} (0.69), V^{3+} (0.74), and Sc^{3+} (0.81) entry are Fe^{3+}

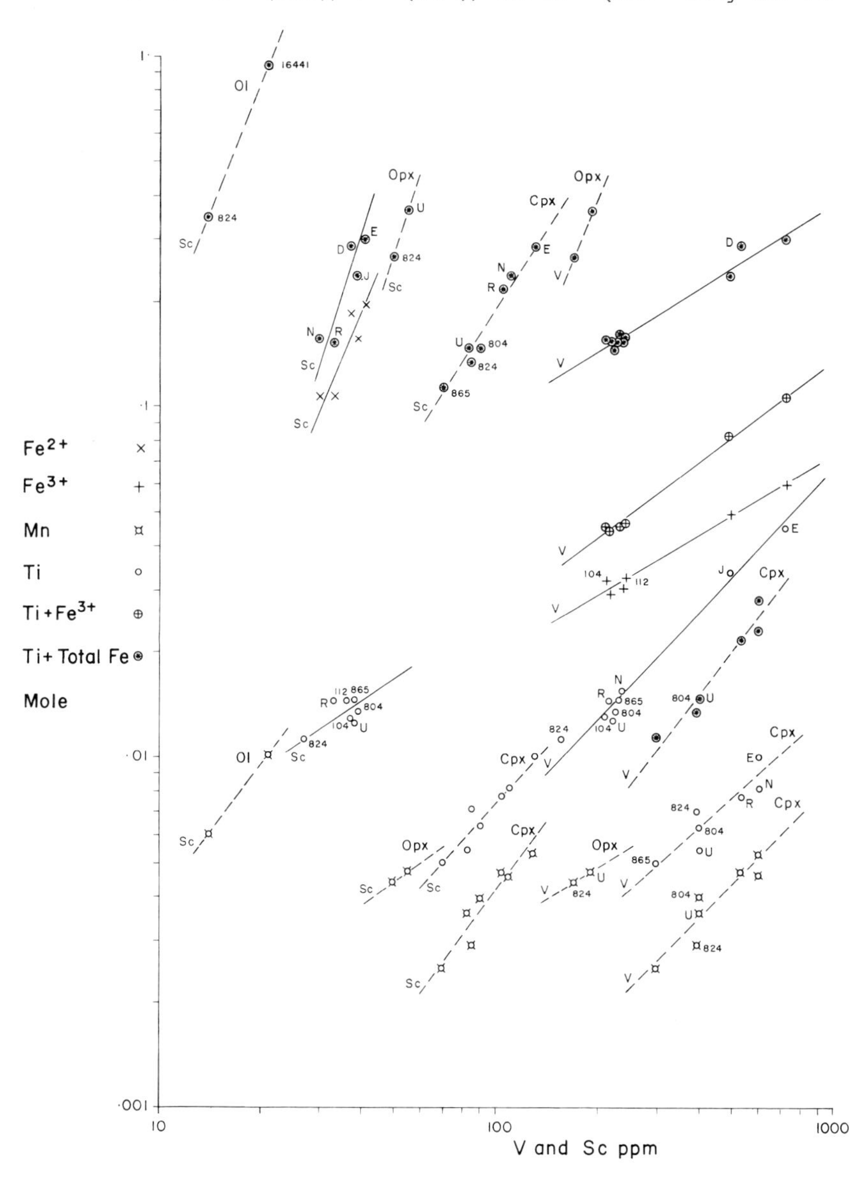

Figure 26. Coherence curves of V and Sc.

(0.64), Ti^{4+} (0.68), Fe^{2+} (0.76), and Mn^{2+} (0.80). These elements are nearly all those in the first set of transition elements. In this group of elements V and Sc provide the best opportunity for the study of element partitioning between the mineral phases and for the study of coherence relationships because they show sufficient variation in concentration with fractionation, and because the magma was not depleted in V and Sc until very late in fractionation. The pattern of entry of Cr, V, and Sc into mineral phases is similar. None of these elements entered plagioclase, other than in traces. All entered apatite and olivine in minor trace amounts only. They all favored entry into the pyroxenes and the opaque iron minerals.

Chromium preference for entry into mineral lattices was clinopyroxene, magnetite, and then orthopyroxene; for V, it was magnetite, clinopyroxene, and then orthopyroxene, whereas for Sc, it was clinopyroxene, orthopyroxene, and then (?) illmenite. Much more Cr entered clinopyroxene than did V or Sc, whereas V entered the opaque iron minerals much more abundantly than did Cr or Sc. From the amounts of Cr, V, and Sc, which entered the clinopyroxene and orthopyroxene (Table 14), it can be seen that about 4 times more Cr entered clinopyroxene than orthopyroxene, whereas the factor was about 2 in favor of clinopyroxene for V, and about 3 to 2 for Sc.

Comparison between Cr, V, and Sc entry into olivine is more difficult, for each entered only in traces, and only Sc provides information on distribution with fractionation. Taking the Mg-olivine layer alone as a guide, the olivine contains about .05, .10, and .25 the amounts of Cr, V, and Sc respectively in the co-existing orthopyroxene. If this is taken as a rough measure of the preference of Cr, V, and Sc entry into Fe^{2+} sites, then they entered in the proportions 1:2:5.

The distribution plots (Figs. 22 to 24) of Cr, V, and Sc indicate that the preferred order of entry into all convenient mineral lattice sites was Cr, V, and Sc, and that Cr had considerable priority over the others. The slopes of the coherence curves also indicate the same order; the low slopes of Cr relationships suggest that it entered appropriate lattice sites most rapidly with fractionation, whereas the high slopes of the Sc curves indicate that its entry was the slowest. The pattern of distribution and coherence shown by V indicates that its behavior is closer to that of Sc than Cr.

These coherence curves may be used in conjunction with the distribution plots to determine the preferred order of entry of Cr, V, Sc, Mn, Ti, and Fe into the minerals of the intrusion with fractionation. The distribution plots, in particular, indicate the order in which the trace elements were removed from the magma with respect to the major and minor elements. The differences in the behavior of the trace elements are determined by the differences in the size of their cations and by the differences in their attraction to, and bonding in, the mineral lattices forming under prevailing physico-chemical conditions in the magma. Their consistent behavior in the mineral series results from the closeness in their chemical properties. The preferred order of entry during crystallization of the intrusion will be similar to that for the mineral series because the main minerals controlling their distribution show insufficient modal variation to affect the order. An analysis of the distribution plots and coherence curves indicates the preferred order is Cr > Mn $\gtreqless$ V > Ti > Sc > Fe.

Burns and Fyfe (1964) have concluded from a study of absorption spectra measurements of transition metal ions in silicate glasses that the order of preferred entry of the ions of these elements into silicate lattices which crystallize from a magma with octahedral and tetrahedral sites can be interpreted according to crystal-field theory in terms of the magnitude of the octahedral "site preference energy." This indicates the following order for M‴ ions:

$$\text{Cr} > \text{Mn} > \text{V} > \text{Ti} > \text{Fe} \geqq \text{Sc}.$$

The preferential entry of Cr over V and Sc into Fe^{3+} sites was apparently assisted by its more favorable ionic size and bonding characteristics in the appropriate lattices available for entry. Similar factors may also have determined the slight priority of V over Sc. The competitive advantage of Cr for Fe^{3+} sites would have relatively reduced both V and Sc entry during early fractionation stages. This limitation was partly overcome by V entering magnetite, which it entered more freely than Cr. But Sc had the additional problems of its relatively large ionic size and of balancing charge if it tended toward the more appropriately sized Fe^{2+} and Mn^{2+} sites. However, the problem of balancing charge upon entry into lattices is common to all three trace elements. They all entered Mn^{2+} positions, and possibly Fe^{2+}. Entry into these lattice positions may have been overcome by Al^{3+} taking the place of Si^{4+} in tetrahedral co-ordination, thus providing an extra charge in the lattice for balancing purposes. Possibly the association between Cr and Al can be attributed to this exchange also, but because of the distinct coherence relationships indicated between these elements in the early clinopyroxenes, the more preferable explanation is that the clinopyroxene structure is favorable for Cr entry into Al sites, whereas in the plagioclase structure such an association was not possible. There appears, however, to be a large discrepancy between the size of their ions, which may indicate that some modification in the size of the Cr ion takes place upon entry.

The valency problem in relation to Cr, V, and Sc entry into Fe^{2+} positions is probably the main cause of their limited entry into the olivine structure, and their reduced entry into orthopyroxene compared to clinopyroxene.

Scandium probably made more use of Fe^{2+} positions than the others, as it not only appears to have some of the weakest bonding characteristics in mineral lattices, but its large cation would also have a restricting influence on entry, and Fe^{2+} and Mn^{2+} appear to offer the largest cation sites in the lattices. However, Sc shows good coherence with Ti, and again size discrepancy in this case appears to be fairly large.

Vanadium and Sc do not appear to have been aided by volatiles, or by excessive element concentration in the magma, to form complexes in the late fractionation stages. Indeed, only traces of V are found in the alkali-enriched parts of the intrusion. Rather, V and Sc entered all convenient lattice sites in competition with Cr, but these were available to V and Sc mainly during late fractionation stages where V had slight preferential entry over Sc, as appropriate early lattice sites were occupied mainly by Cr.

Chromium, V, and Sc probably entered Fe sites in plagioclase, and are present in only minor trace amounts. Apparently the plagioclase structure did not provide convenient entry. Similar trace amounts are in apatite, where probably Fe or possibly Mn sites offered the most favorable entry.

Cobalt and Nickel

Cobalt. The original intrusion contained 53 ppm Co and the average for the intrusion is 56 ppm. As shown in Figure 27, the distribution of Co in the intrusion has been strongly influenced by the distribution of hyalosiderite in the early fractionation stages. Excluding the Mg-olivine layer, where Co has a peak concentration of 140 ppm, Co tended to build up in the magma

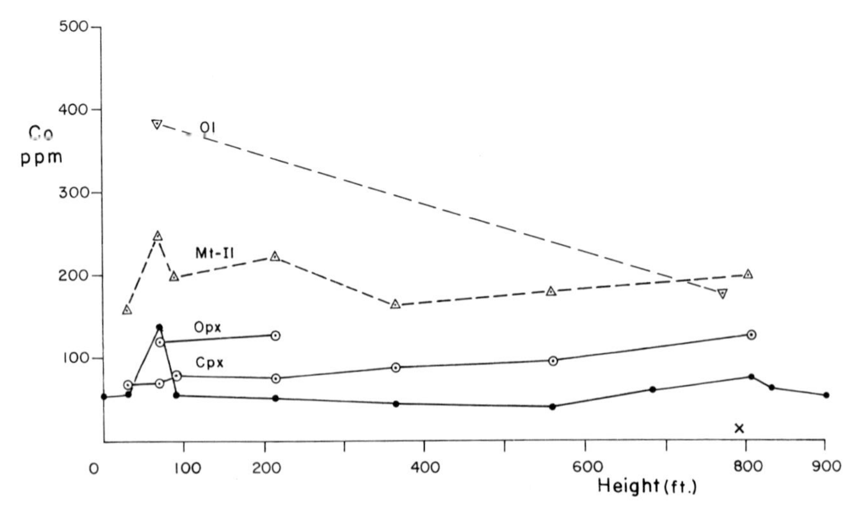

Figure 27. Distribution trends of Co, Englewood Cliff Section.

during the early and middle fractionation stages, and entered late-stage rocks, of which the ferrodolerites contain up to 75 ppm. The magma was depleted in Co after this stage, and values decreased to 27 ppm in the fayalite granophyre and the granophyric dolerite. The content in rocks and minerals is given in Tables 13 to 16.

Cobalt partitioning between minerals was dominated by olivine in the early stages. However, fayalite was not the most favorable location in the late stages, so that Co decreases in the olivine series with fractionation. The reverse trend is shown in the opaque iron mineral, orthopyroxene, and clinopyroxene mineral series. Of these, Co preferentially entered the opaques, whereas orthopyroxene, and in turn, clinopyroxene, had less favorable lattice sites. Cobalt was not detected in plagioclase, and it entered apatite in minor trace amounts only.

In the mineral series Co coherence is mainly with Mg in the olivines, and with Fe^{2+} in the orthopyroxenes and clinopyroxenes (Fig. 29). However, in the late fractionation stages the Co coherence in the clinopyroxenes appears to be entirely with Fe^{2+}, whereas in the early stages it is with (Fe^{2+} + Mg). That is, the olivine lattice favors entry of Co into sites normally occupied by Mg, whereas the orthopyroxene and clinopyroxene crystal lattices favor Co entry into sites normally occupied by Fe^{2+}.

Cobalt distribution (and also that of Ni) in the opaque iron minerals closely parallels that in the whole rocks. They apparently follow Fe^{2+}. However, if the opaques analyzed are dominantly magnetite, the iron oxide structure could also contain some Mg.

In the Palisades intrusion, the modal composition of olivine, orthopyroxene, and clinopyroxene in the whole rocks determine Co coherence with Mg in early fractionation stages. But in the late stages the modal composition is dominated by clinopyroxene and the opaque iron minerals, and consequently Co shows coherence with Fe^{2+}. In the middle stages coherence is, as expected, with (Fe^{2+} + Mg). The amount of Co and Ni in the opaques has only a small influence on the coherence relationships observed in the whole rocks because the distribution trends of Co and Ni parallel those in the whole rocks, and because the modal contribution of the opaques is small in the early fractionation stages.

Nickel. In contrast to Co, nickel was removed rapidly from the magma in the early fractionation stages. Nickel in the chilled dolerite is 95 ppm, but its concentration rises sharply in the early fractionation stages to a peak of 500 ppm in the center of the Mg-olivine layer. Just as sharply it falls to 135 ppm in the bronzite dolerite immediately above the layer, and subsequent dolerites of the fractionation series contain progressively less Ni; the ferrodolerites average 30 ppm and the fayalite granophyre contains less than 10 ppm. The average Ni content of the intrusion is 85 ppm. The Ni content of rocks and minerals is given in Tables 13 to 16, and the distribution of Ni with fractionation is shown in Figure 28.

Nickel concentrated in the Mg-olivine layer, where it is mainly in the olivine and opaque iron minerals, and in lesser amounts in orthopyroxene and clinopyroxene. The early entry of Ni into mineral lattices was not as complete as it was for Cr, so that the magma retained some Ni for distribution throughout fractionation, though at much reduced concentration levels in the late fractionation stages. The magma was almost entirely depleted in Ni by the fayalite granophyre stage, but significantly more Ni entered the low temperature clinopyroxene than the co-existing fayalite, though in the Mg-olivine layer the hyalosiderite provided much more favorable entry for Ni than did the co-existing pyroxene.

The order of preferred entry into ferromagnesian minerals was olivine, orthopyroxene, and clinopyroxene. Excluding the effect of the Mg-olivine layer on Ni distribution, Ni content in each mineral series decreases with fractionation. The trend of Ni in the clinopyroxene is similar to that in the whole rock. Only traces of Ni occur in plagioclase and apatite.

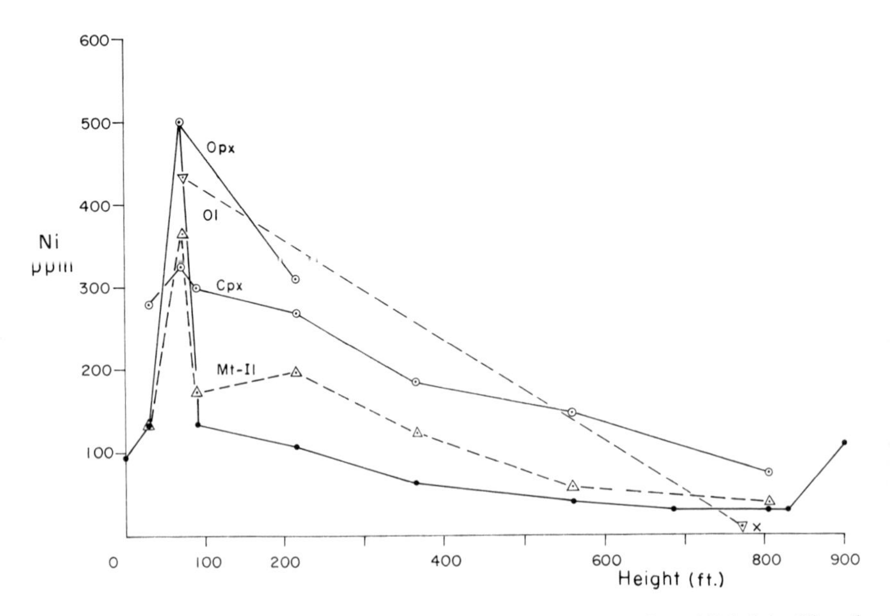

Figure 28. Distribution trends of Ni, Englewood Cliff Section. Nickel in Ol and Mt-Il shown at one-third scale.

In all the ferromagnesian mineral series, Ni coherence is mainly with Mg (Fig. 30), and thus most Ni in the late fractionation stages entered the mineral lattices with the most Mg, and not the most Fe^{2+} sites. The augite in the Mg-olivine layer apparently has the highest Mg content of the clinopyroxenes in the intrusion because it has the highest Ni content. In the whole rocks, Ni coherence is with (Mg + Fe^{2+}) in the early fractionation stages, and in the middle and late stages it is with Mg.

The possibility that minute inclusions in some pyroxenes separated for analyses are a mineral phase high in Cr has been referred to. Similarly, such a phase could contain Ni and Co, but it does not appear that this possibility affects the conclusions reached on the geochemical behavior of either Ni or Co.

Geochemical behavior. The closeness in the chemical properties of Co, Ni, Fe, and Mg results in the very similar behavior shown by Co and Ni. Cobalt and Ni preference of entry into minerals is olivine, opaque iron mineral, orthopyroxene, and clinopyroxene, and in the Mg-olivine layer the Ni/Co ratio in these minerals is approximately 4 to 1 (Tables 14 and 16). The distribution trends of each with fractionation are complementary, and the behavior of Co is the better for element partition studies for it maintains coherence relationships in all but the very late-stage fractionation products. By contrast Ni was largely removed from the magma in early fractionation stages. Though Co and Ni show many similarities in their behavior, they discriminate between Fe^{2+} and Mg sites according to lattice type; preference of entry appears to be determined almost solely by the order of increasing magnitude in the bond energies of the cations in the different lattice structures, as other properties controlling entry appear to be nearly equal.

Many of the theoretical explanations of the geochemical behavior of Co and Ni to date have been inconclusive, and because of apparent inconsistencies in observed behavior, it has been implied that their behavior is complicated.

Ringwood (1955a) predicted on the basis of electronegativity differences that Ni substitutes for early Fe^{2+} rather than for Mg. This substitution, he claimed, would also be assisted by the more ionic character of the Mg-O bond compared with that of Ni-O, which, in turn, he claimed is stronger than the Fe-O bond. Previously, however, Goldschmidt (1954), following largely the observations of Vogt (1923) on the distribution of Ni in natural occurrences, explained Ni camouflage for Mg on the basis that ionic radii and bonding favored this substitution.

Taylor (1966), in summarizing the status of the problem, indicated that the convenient size of Ni^{2+} (0.72) and Co^{2+} (0.74) between Mg^{2+} (0.65) and Fe^{2+} (0.76), together with melting-point data for the oxides, would

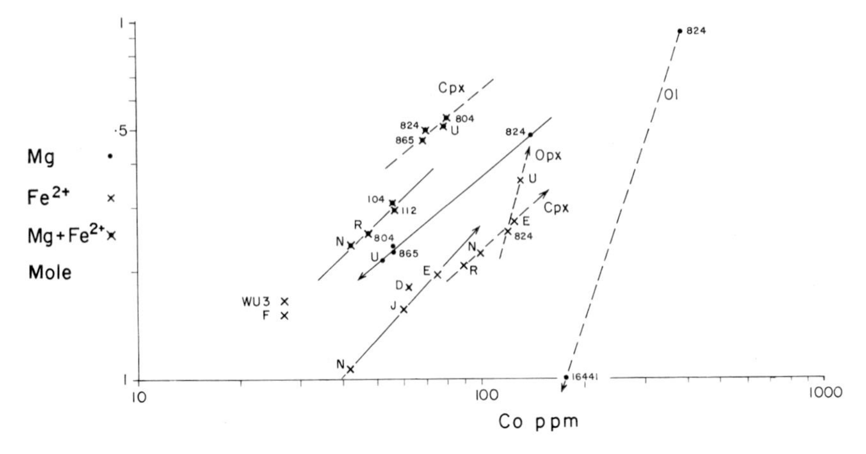

Figure 29. Coherence curves of Co.

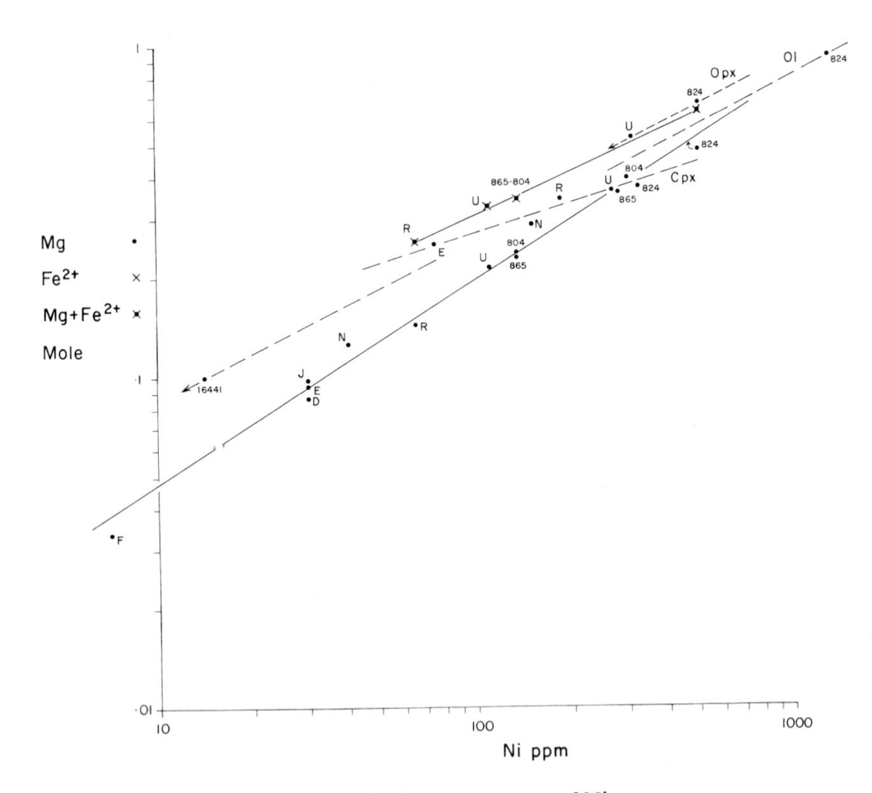

Figure 30. Coherence curves of Ni.

suggest that the sequence of entry into crystal lattices would be $Mg^{2}+$, $Ni^{2}+$, Co^{2+}, Fe^{2+}, particularly as there are no charge balance problems. However, he was unable to explain the more favorable entry of Ni than Co into early ferromagnesian mineral lattices. He believed that both Ni and Co occupied the Fe^{2+} sites.

Burns and Fyfe (1966) point out that melting-point data alone are not necessarily a guide to bond energies, nor to the order of preference that an element will show in incorporation in natural basaltic systems. But greater confidence in prediction can be placed in lattice energies, which, for example, indicate that the Ni-O bond energy is larger than those of Mg-O and Fe-O in the solid phase. These observations appear consistent with the behavior of Ni shown in the Palisades Sill, and probably crystal-field theory provides the most satisfactory explanation of its behavior.

In the Palisades, Ni and Co enter either Mg or Fe^{2+} sites according to the lattice characteristics of the mineral, and the bond energies of the cations apparently vary somewhat between structures. An analysis of the distribution trends of Ni and Co indicates that Ni had preferred entry over Co into suitable lattice sites, particularly in the early fractionation stages. Furthermore, an analysis of the coherence curves (Figs. 29 and 30) also shows that Ni entered mineral lattice sites much more rapidly than Co with fractionation because the atomic weights of both are very close and the curves of Ni are of much lower slope than the corresponding curves of Co. In olivine both Ni and Co entered Mg sites, but Ni had priority over Co so that the preferred order of entry is Ni $>$ Co $>$ Mg $>$ Fe^{2+}. In clinopyroxene and orthopyroxene, Ni preferred Mg sites, and Co preferred Fe^{2+}, and the preferred order of entry is Ni $>$ Mg $>$ Co $>$ Fe^{2+}. Thus, it appears that the bond energy of Mg in olivine is weaker than that of Co, whereas the reverse is the case in the pyroxene. For the whole rocks the sequence of entry will vary between intrusions according to the modal distribution of minerals containing Co and Ni.

The coherence curve for Ni in the whole rocks indicates that Ni may follow Fe^{2+} in part in the early fractionation stages, but that coherence with Mg became progressively more dominant with fractionation. Relationships in the mineral series show that Ni follows Mg in the ferromagnesian minerals. However, it is probable that it follows Fe^{2+} in the opaque iron minerals. The relationship with Fe^{2+} in the whole rocks could be apparent rather than real, but it is possible that some entry into Fe^{2+} sites took place during the stage of rapid removal of Ni from the magma, and this may be in some way connected with the degree of ionic activity or with the relative ion concentrations of Ni, Mg, and Fe in the liquid in the early cooling stages of the intrusion. Nickel preference for Mg sites is only marginally more favorable than it is for Fe^{2+} in the ferromagnesian mineral lattices, though the distribution plots demonstrate that Ni has much greater priority of entry into lattice sites than either Mg or Fe^{2+}. Nickel can enter either conveniently, and entry into both might be expected in a mineral such as olivine, whose Fe^{2+} and Mg sites are presumably identical.

The minor trace amounts of Ni in plagioclase are probably present in either Fe^{2+} or Mg sites; however, the plagioclase crystal structure did not provide convenient entry for either Ni or Co. Both entered apatite in minor trace amounts, and, again they probably entered the Fe^{2+} sites.

The results obtained for Ni and Co distribution and partitioning in the Palisades Sill illustrate how necessary it is to evaluate coherence relationships in the mineral series before interpreting those in the whole rocks. The coherence of Co and Ni with a particular major element in the whole rocks with fractionation depends on the modal distribution of the minerals in which they occur, and on the coherence they show with either Mg or Fe^{2+} in these minerals. As mineral composition and abundance in the rock vary with fractionation, so will the coherence of Co and Ni with Mg and Fe^{2+}.

Observations on Ni and Co behavior in other intrusions also appear consistent with these results. For example, Carr and Turekian (1961) maintained that Co showed good coherence with ($Mg + Fe^{2+}$), rather than with either Fe^{2+} or Mg independently. This conclusion was reached from an empirical study of Co in basic igneous rocks in general, without regard to the fractionation stage that each represented. The world-wide collection of basic rocks includes sufficient numbers of both early- and late-stage varieties for the representative sampling jointly to show a coherence between Co and ($Mg + Fe^{2+}$).

McDougall and Lovering (1963) indicated that in the Red Hill intrusion Co shows coherence with Fe^{2+} in the whole rocks. Again this is fairly consistent with observations on the whole rocks of the Palisades intrusion, for the fractionation range in Red Hill corresponds roughly to the middle and late fractionation stages in the Palisades.

Yttrium, Zirconium, Niobium, and Molybdenum

Yttrium. Yttrium concentrated in the magma during the early and middle fractionation stages and was removed during late stages. The original magma contained 29 ppm Y, which is about the same as the average for the intrusion. The distribution trends of Y are shown in Figure 31, and values in rocks and minerals are given in Tables 13 to 16. The trend in the whole rocks is roughly similar to that of Ba.

Yttrium entered olivine, clinopyroxene, and orthopyroxene in roughly equal amounts. It is more than twice as abundant in the opaque iron mineral

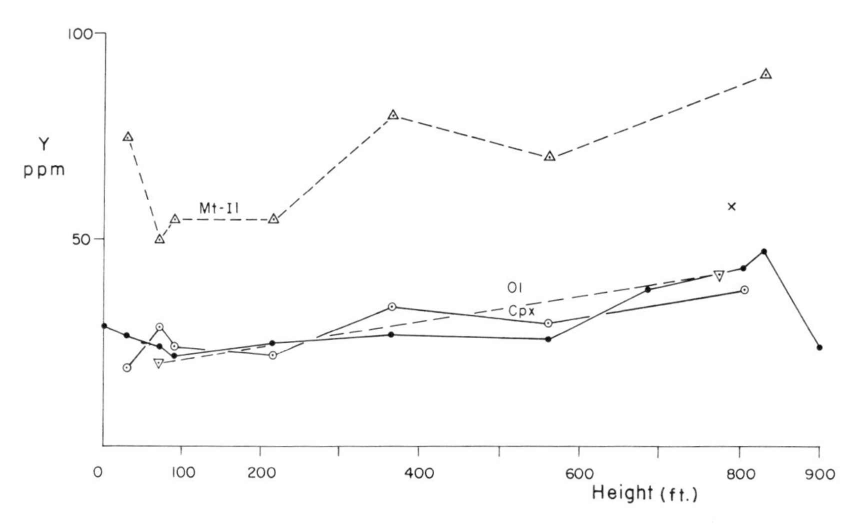

Figure 31. Distribution trends of Y, Englewood Cliff Section.

as it is in the ferromagnesian minerals. In the apatite of the fayalite granophyre, 3600 ppm are present. Indeed, apatite appears to have provided a very favorable location for both Y and La, and probably was also the main collector of the rare earth elements in the intrusion. Unlike Ba, however, yttrium entered plagioclase only in traces. The readiness with which Y associates with apatite, and probably also with the other accessory minerals—for example sphene—makes the interpretation of its behavior with fractionation difficult to follow as insufficient analytical data are available on these minerals.

On the basis of ionic properties, it has been argued that Y^{3+} (0.93) should enter Ca^{2+} (0.99) because of its somewhat smaller size and valency advantage. However, Taylor (1966) indicates that because Y normally concentrates in the late-stage fractionation products, entry may be restricted by its more covalent bond character as compared to that of Ca-O. The distribution of Y in the rocks and minerals of the Palisades intrusion shows that, except possibly for apatite, the Y seems to have entered into Ca-bearing minerals with difficulty. No coherence is shown between Y and Ca in the clinopyroxene series; in fact, an inverse relationship is apparent between them. The limited entry of Y in similar amounts into clinopyroxene and olivine also suggests that Ca was not the favored lattice site.

The coherence of Y^{3+} in the minerals and whole rocks is with Ti^{4+} (0.68), Fe^{3+} (0.64), and Fe^{2+} (0.76). The relationships are shown in Figure 32. Yttrium shows good coherence with Ti in the clinopyroxene series, and

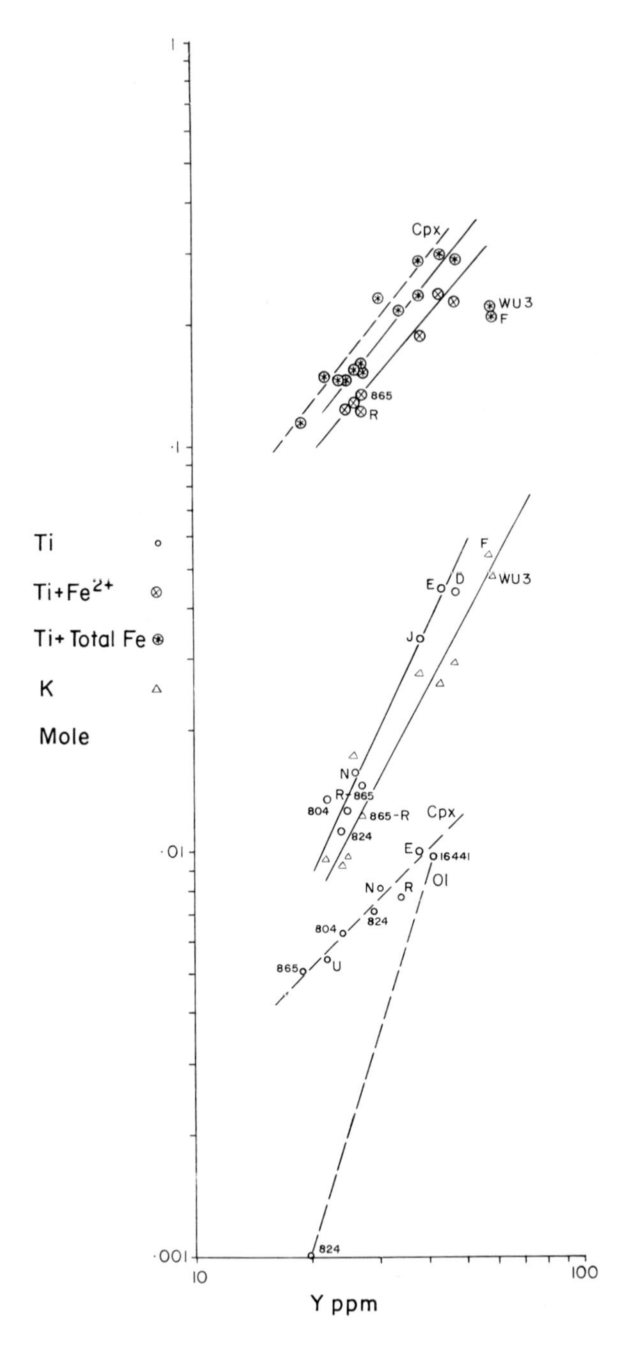

Figure 32. Coherence curves of Y.

some coherence with Ti is indicated in the olivine and opaque iron mineral series. The discrepancy in ion size is large, but apparently lattice properties of the minerals concerned tolerate this discrepancy, and the other chemical properties determine Y entry into sites normally occupied by Ti. Also in the clinopyroxene series Y follows Fe, but the slopes of the coherence curves suggest that it did not enter Fe quite as easily as Ti sites with fractionation. The Y with Fe coherence relationship in whole rocks indicates that Fe^{2+} and Fe^{3+} sites provided equally favorable entry for Y. In either case, but particularly for Fe^{3+}, ion size differences are large, and though size favors entry into Fe^{2+} sites, valency difference might have countered this advantage.

In the whole rocks a relationship between Y and (Ti + Fe) is indicated, but that with total Fe in the middle and late fractionation stages appears the more pronounced according to the coherence curve slopes. The relationship with Ti persists throughout fractionation, except in the very late stages. That with total Fe fails, both in the very early (where the Mg-olivine layer is anomalous, as it is also for Y coherence with Fe in the clinopyroxene series) and in the very late fractionation stages. In addition to the Y and (Ti + Fe) relationship there is another between Y and K, but in this case the coherence appears somewhat better, as the plots of the late-stage granophyric rocks lie on the curve. It is not possible to verify this relationship in any mineral series at present, so it could be apparent rather than real. Yttrium, however, appears to have stronger bonding characteristics than either Ti or K.

The behavior of Y in the very late fractionation stages is not clear without further analytical work on the accessory minerals, as the coherence with Ti and Fe lapses in the fayalite granophyre and the granophyric dolerite. Various explanations can be advanced. As already mentioned, Y appears to have entered apatite freely. Apatite occurs throughout the intrusion, as a very minor accessory in the early and middle fractionation stages, and as an important accessory in the late stages. It may be argued then that the entry of abundant Y into apatite in very late stages caused the coherence relationship with Ti and Fe to lapse. But cessation of this relationship could also be due to depletion of the magma in Y.

In the Y association with apatite, Y must either enter Ca sites, or form a separate phosphate phase with apatite, or, less likely, it is aided in its entry by halogens, which are common in some apatites. Lieberman (1966) determined the Br content of the type specimens, and found Br present in minor trace amounts. The distribution trend of Br is similar to that of Y and particularly to that of La (Fig. 34).[5] Entry of Y into Ca sites has been considered

[5]Bromine probably enters apatite. If the variation of P concentration with fractionation can be taken as a guide to the modal distribution of apatite in the intrusion, then it suggests that Br enters apatite because the distribution trends of Br and P are similar.

reasonable by Beevers and McIntyre (1945) for, they explain, Ca is present in 9-fold co-ordination with oxygen and has large lattice sites. Also attractive, however, is the possiblity of the formation of a separate phosphate phase intergrown with apatite, though no petrographic evidence has been found in support of this.

Zirconium. The distribution trend of Zr in the whole rocks indicates that Zr concentrated slightly in the magma in early fractionation stages, and crystallized mainly during late stages. The trends in rocks and minerals are shown in Figure 33, and the values are presented in Tables 13 to 16. The original magma contained 120 ppm Zr, and the average composition of the intrusion is about 130 ppm. The highest concentrations in the whole rocks occur in products of residual fractionation; the granophyric dolerite contains 330 ppm.

The main feature of Zr behavior is that it forms its own silicate phase, zircon, rather than enter other mineral lattices. This accounts for its somewhat irregular distribution in the whole rocks. It did, however, consistently enter the opaque iron minerals in amounts slightly less than 300 ppm. Furthermore, 150 ppm Zr was found in the fayalite, and 200 ppm in the apatite. But it is below 50 ppm, the detection limit of the analytical method used, in the pyroxenes, plagioclases, and Mg-olivine.

It appears that in the Palisades intrusion Zr avoided all Ca-bearing minerals, except apatite. In the opaque iron minerals it shows a gentle but gradual increase in amount with fractionation, similar to the Zr trend in the

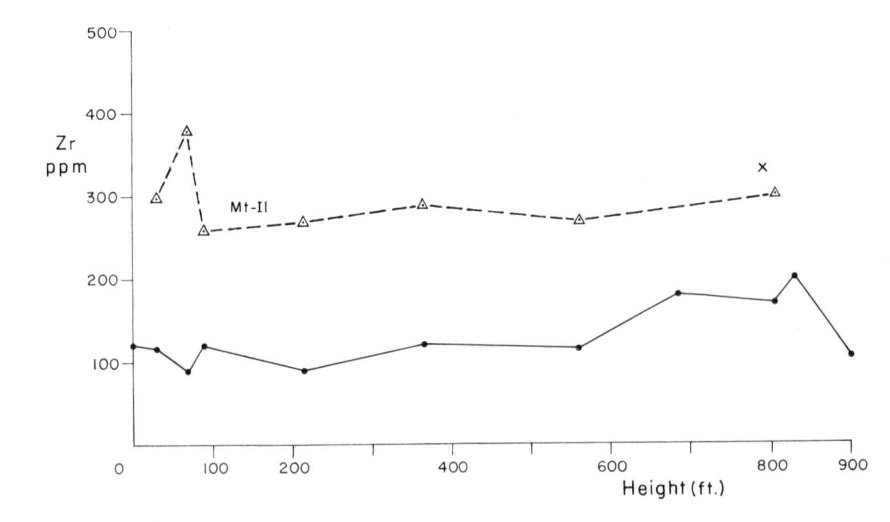

Figure 33. Distribution trends of Zr, Englewood Cliff Section.

whole rocks. In the Mg-olivine layer, however, Zr rises to 380 ppm in the opaques, whereas in the whole rock it falls to 90 ppm. The opaques account for only 1 ppm of the Zr in the hyalosiderite dolerite, so the remainder must be zircon, as Zr has not been detected in the other minerals. This pattern of Zr distribution is probably consistent throughout the intrusion. The occurrence of Zr as a separate mineral phase in quantities too small to measure modally makes the complete evaluation of Zr behavior difficult.

Thus, the main feature of Zr distribution in the intrusion is that most of the Zr did not behave like a normal trace element. Indeed, its behavior was more like that of a major element in that its concentration built up in the magma until it formed an independent silicate phase; it was not readily accommodated in other mineral lattice sites. Distribution of Zr in the whole rocks indicates a rough coherence relationship with Y and La, and possibly Y could, in certain circumstances, be expected to follow Zr in its distribution as Y appears to form the stronger bond, but would have size and charge balancing problems.

The less significant facet of Zr behavior is its entry into lattice sites of other minerals. The size of Zr^{4+} (0.80), in addition to the problem of charge balance, made entry into major cations sites of minerals difficult. Ringwood (1955b) has indicated that Zr may also form complexes. As $(ZrO_4)^{4-}$ is large with respect to $(SiO_4)^{4-}$, it concentrates in residual melts. The closest major cations in size to Zr^{4+} are Fe^{2+} (0.76), Ti^{4+} (0.68), and Fe^{3+} (0.64). There is a close similarity in the distribution of Zr and La in the whole rocks, and a rough similarity to Y and Sc distributions. Both Y^{3+} (0.93) and Sc^{3+} (0.81) have a similar size problem with respect to Ti and Fe, though entry into these lattice sites appears possible in the pyroxenes and the opaque iron minerals. Possibly, therefore, some entry of Zr into these minerals also could be expected. Wager and Mitchell (1951) observed that Zr entered the Skaergaard pyroxenes and that the Zr content declined with fractionation while the Zr in the magma was rising strongly.

The distribution trends of Zr and Ti are similar in the early and middle fractionation stages of the Palisades intrusion, though much less so in late stages, where undoubtedly most of the Zr crystallized as zircon from the magma. Taylor (1966) has suggested that entry of Zr^{4+} into Ti^{4+} sites is logical in that the Zr-O bond appears to be the more ionic. No clear pattern emerged in the Palisades rocks from the distribution of Zr and Ti in the opaque iron minerals, which contain abundant Ti relative to Zr. Probably Zr in fayalite also follows Ti, as very little Ti is present in the hyalosiderite.

It has been suggested by Wager and Mitchell (1951) that possibly Zr enters Ca sites in apatite. However, if Y and La form phosphate phases with apatite, Zr would be more likely to follow them, but its entry would be limited as it appears to have weaker bonding characteristics.

Niobium. Niobium was detected only in trace amounts in products of late fractionation stages, and is probably a relatively unimportant element in the fractionation of the Palisades magma. The various mineral phases have not been analyzed for Nb. The exploratory analysis indicated that Nb was retained in the magma during much of the fractionation, and concentrated in residual melts.

Molybdenum. Molybdenum in the whole rocks was detected only in the fayalite granophyre where it is present in amounts slightly above the detection limit of the analytical method used. Molybdenum mostly entered the opaque iron minerals, and its presence in these shows a progressive, but irregular, increase in concentration with fractionation (Fig. 37). It was also determined in the olivines, where again concentration increases with fractionation. Only traces are present in plagioclase, and 15 ppm was found in the apatite analyzed. It was not detected in pyroxene. Values are given in Tables 13 to 16.

Molybdenum shows a strong tendency to complex and concentrate in residual magmas, as the complexes formed are larger than $(SiO_4)^{4-}$ (Ringwood, 1955b). Taylor (1966) suggests that the free Mo^{4+} (0.70) ion can enter Ti^{4+} (0.68), Zr^{4+} (0.80), or Fe^{3+} (0.64), which is consistent with its presence mainly in the opaque iron minerals. In olivine and apatite it apparently followed Ti and Fe.

Lanthanum and Neodymium

Lanthanum. The distribution trend of La in the whole rocks is similar to that of Sr and Ti in the early fractionation stages, and Ba and Y in the late stages. The trends of La in the rocks and minerals are shown in Figure 34, and the values are given in Tables 13 to 16. The original magma contained 30 ppm La, whereas the average content in the intrusion is 27 ppm.

Lanthanum concentrated mainly in apatite, and the apatite in the fayalite granophyre contains 3800 ppm. About twice as much La entered the opaque iron minerals as Y, whereas the relative concentrations are reversed in clinopyroxene, orthopyroxene, and olivine, and La is about half the concentration of Y. Many La values for the ferromagnesian minerals border the detection limit (14 ppm) of the analytical method used. Lanthanum was not detected in plagioclase.

The common cations closest in size to those of La and the elements in the rare earth group are Ca^{2+} (0.99) and K^{+} (1.33). Ringwood (1955a) indicated that the large size of La^{3+} (1.15) and Nd^{3+} (1.08) relative to Ca^{2+}, together with the more covalent character of the RE-O bond, lead to their concentration in residual melts. However, La has a valency advantage over

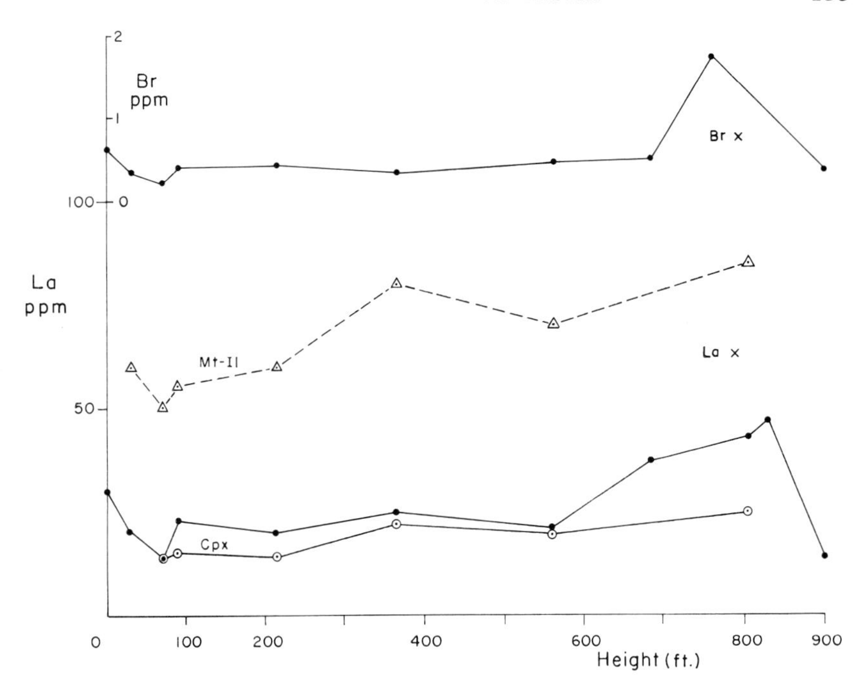

Figure 34. Distributions trends of La and Br, Englewood Cliff Section. Lanthanum in Mt-Il shown at half scale.

Ca, and this may balance, somewhat, its apparently weaker bonding characteristics.

The study of element distributions in the mineral series (Fig. 35) suggests that in the clinopyroxene series La follows Ca and also (Ti + total Fe) in the early fractionation stages. In the later stages the coherence is clearly with (Ti + total Fe) only. The lack of clarity in the relationship with Ti in the early stages probably results from the La values being near the detection limit of the analytical method used. Lanthanum in the opaque iron minerals shows a rough coherence with Ti, and, had Fe values been available, a similar coherence with total Fe could be expected.

Lanthanum distribution in the whole rocks indicates a relationship with (Ca + K) in the early and middle fractionation stages, though control of the curve is poor. The relationship in the middle and late stages is with (Ti + total Fe). In the very late stages this relationship lapses, and this may result, as in the case of Y, from late depletion of the magma in La, or in part, at least, from the increasing importance of apatite in the control of La distribution. The lapse in the relationship roughly coincides with the increase in the

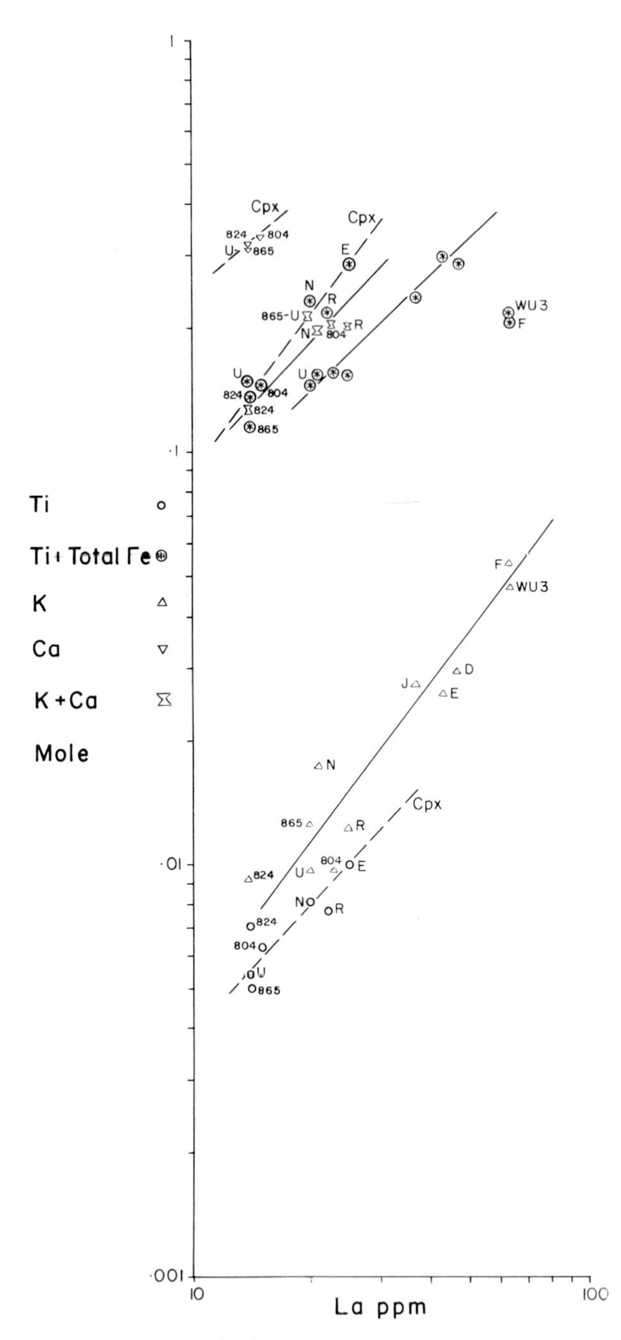

Figure 35. Coherence curves of La.

abundance of apatite in the intrusion. As for Y, the entry of La into apatite may in some way be occluded in mixed phosphate formation.

Throughout fractionation a rough coherence relationship is apparent between La and K in the whole rocks. This coherence has not been tested in any mineral series to date, but a relationship with K could exist as Ba (A.W. 137), a neighboring element to La (A.W. 139) in the Periodic classification, shows a similar relationship. The behavior of La in olivine could be similar also to that of Ba.

The slopes of coherence curves do not indicate any obvious priority for La entry into the major cation sites of minerals; but lower slopes, together with the much higher atomic weight, compared to those of Y, suggest that La entered lattice sites somewhat more easily than Y with fractionation where they are competing for the same positions. Neither element, however, entered silicate mineral lattice sites conveniently, and both, as a result of their size and bonding characteristics, tended to build up concentration in the magma with fractionation.

Neodymium. Only brief comment is possible on Nd, which would be expected to behave similarly to La. Concentrations of Nd in the rocks border the detection limit of the analytical method used. It is present in minor trace amounts only, except in the granophyric dolerite, which contains 35 ppm (Table 13). It was not detected in the minerals analyzed, but it was not analyzed for in apatite, where it is most likely to occur. Indications are that, during fractionation, the magma was enriched in Nd, and that Nd crystallized in products of late-stage fractionation.

Copper

Most of the Cu in the intrusion crystallized as sulfide phases, mainly in late fractionation stages, and distribution trends are somewhat irregular because of this behavior. Copper also entered all mineral phases in small amounts. Its distribution in rocks and minerals is shown in Figure 36, and values are given in Tables 13 to 16, where it can be seen that the original magma contained 110 ppm Cu. The average Cu composition of the intrusion is 135 ppm. In the first part of the early fractionation stages, Cu in the whole rocks falls progressively to 90 ppm in the Mg-olivine layer and the early dolerite immediately above. Thereafter, it shows a slight but steady increase to 120 ppm in the late pigeonite dolerite stage. Copper precipitated from the magma mainly during the late stages, and reached a maximum concentration of 300 ppm in the granophyric dolerite; it is also abundant in the ferrodolerite, which contains 250 ppm. Thus Cu tended to remain in the magma until late stages, apparently until its concentration built up sufficiently for an independent sulfide phase to form. It is of interest, too, that Cu is one of the

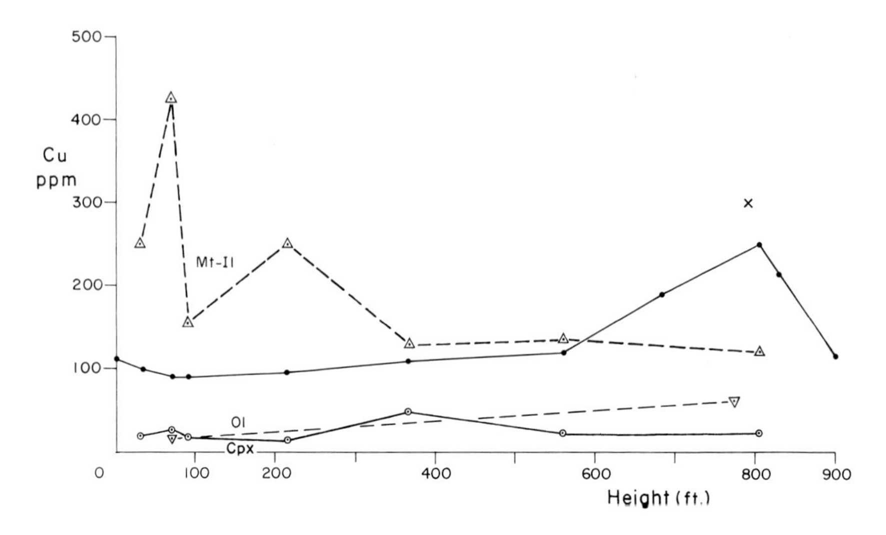

Figure 36. Distribution curves of Cu, Englewood Cliff Section.

few elements whose distribution was unaffected by the formation of the Mg-olivine layer, indicating that it is largely independent of silicate mineral lattices at this stage. However, some entry into crystal lattices is indicated by the slight decrease in the plagioclase trend, and the slight increase in the pyroxene trend, in the layer.

Distribution of Cu in mineral phases appears unrelated to Cu distribution in the whole rocks; the small amounts in the olivines, clinopyroxenes, plagioclases, and apatite probably include Cu in the mineral lattices, but may also include copper as minute grains of sulfide, or Cu, adsorbed to crystal surfaces along cleavages and walls of fine crystal fractures.

Examination of polished sections indicated that Cu occurs throughout the intrusion as independent small chalcopyrite grains in association with the opaque iron minerals. Disseminated grains are most common in late fractionation stages and toward the top of the intrusion in the late hydrothermal veins; calcite-rich veins also contain chalcopyrite. Copper content in the opaque iron minerals of the Mg-olivine layer is 425 ppm, which is nearly as high as it is in the opaques of the fayalite granophyre, which contains 525 ppm.

The Cu^{2+} (0.72) ion is closest in size to Fe^{2+} (0.76) whereas Cu^{+} (0.96) is similar in size to Na^{+} (0.95). Ringwood (1955a) indicates Cu-O forms a weaker and more covalent bond than either Na-O or Fe-O. He concludes that Cu does not readily enter silicate minerals, though it does so to a limited extent, and is found in plagioclase in Na^{+} positions and in the ferromagnesian minerals in Fe^{2+} sites. In apatite and the opaque iron minerals, it also probably followed Fe^{2+}.

The dominant characteristic of behavior of Cu is its tendency to complex in the magma rather than form independent ions. When sufficiently concentrated Cu forms an immiscible liquid, which in turn separates as a sulfide phase. This behavior was identified in the Skaergaard intrusion by Wager and Mitchell (1951). Here, they also found that there is a sharp drop in the Cu content of the silicate minerals where heavy sulfide precipitation first occurs, though Cu in the whole rock continues high.

Copper behavior in the Palisades is consistent with that in the Skaergaard, except that values for Cu in silicate mineral phases are much lower in the Palisades, and, in particular, in plagioclase. McDougall and Lovering (1963) found that Cu behaves similarly in the Red Hill intrusion.

Boron and Gallium

Boron. Boron concentrated in residual melts, and is present above the detection limit of 20 ppm only in rocks of late fractionation stages. Traces occur in the earlier fractionation stages, and in the chilled contact dolerite. Rocks of the most volatile-rich stages are the richest in B. The pegmatitic dolerite contains 70 ppm and the granophyric dolerite 140 ppm. Distribution of B in the whole rocks is shown in Figure 37 and values are given in Table 13.

Ringwood (1955b) has indicated that B^{3+} (0.20) ions do not exist in the magma. The B^{3+} ion is less than half the size of Al^{3+}, and there are no other chemically similar major cations for B to follow. Probably B occurs as $(BO_3)^{3-}$, which is not readily accepted into structural positions in silicates, and thus it tends to concentrate in residual melts. However, the tendency for B to form complexes also leads to $(BO_4)^{5-}$, which can take the place of $(AlO_4)^{5-}$ in early-formed minerals because of its smaller size.

Friedman (1954) estimated the concentration of B in the upper and lower contact dolerites of the Palisades, and found that it is more abundant in the upper. In considering the possible significance of this, he speculated on the addition of B to the magma through contamination by overlying sediments. As already noted, field evidence shows that assimilation, where present, is superficial; the likely explanation is that the B fractionated to higher levels in the intrusion during crystallization.

Gallium. The Ga content of the magma does not vary much with fractionation. The composition of the original magma is 14 ppm, whereas the average for the intrusion is 17 ppm. Concentration in the whole rocks is virtually unchanged in the Mg-olivine layer, and, with fractionation, the content in the rocks built up to about 20 ppm in the late pigeonite dolerite,

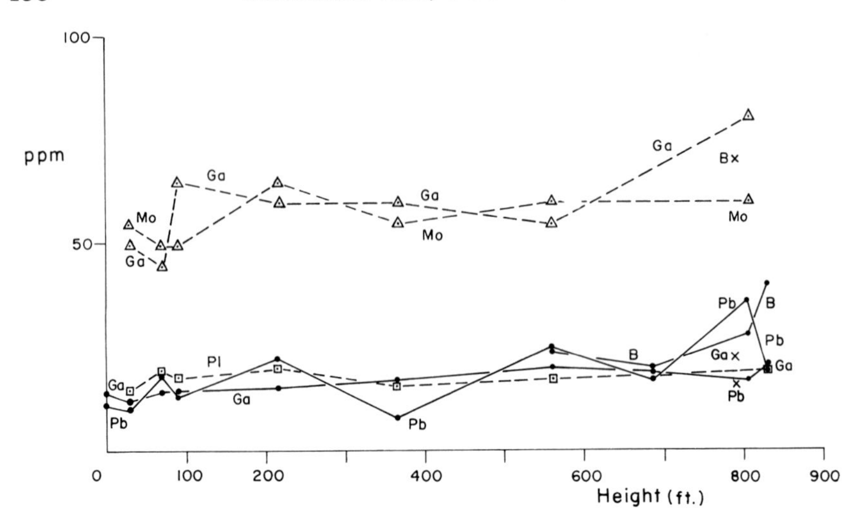

Figure 37. Distribution trends of Ga, Pb, Mo, and B, Englewood Cliff Section. Boron in rock W-F-60 (BX) shown at half scale.

ferrodolerite, and fayalite granophyre, and reached a maximum of 22 ppm in the granophyric dolerite. Distribution trends of Ga in the whole rocks and minerals are shown in Figure 37, and values are given in Tables 13, 15, and 16.

The distribution trend of Ga in plagioclase is similar to that in the whole rocks, whereas that in the opaque iron minerals is somewhat irregular. Gallium content in the plagioclase series increases slightly in the Mg-olivine layer, and this contrasts with the rather sharp decrease in the opaque iron mineral series. Although the Ga content of the plagioclase is only about one-third that of the opaque iron minerals, plagioclase is mainly responsible for the Ga content of the rocks, as plagioclase is modally the much more abundant mineral. Gallium was not detected in pyroxene, olivine, or apatite; however, the analytical method used was not very sensitive for Ga determination. For this reason the plagioclase distribution trend for Ga is shown by a broken line in Figure 37.

Though the range in Ga composition in the whole rocks and in each mineral series throughout the intrusion is small, Ga shows rough coherence in its distribution with Al, a relationship that Wager and Mitchell (1951) and Nockolds and Allen (1956) have noted previously in other intrusions. In the Skaergaard, Ga entered plagioclase and magnetite fairly readily, but occurs only in minor amounts in pyroxene, olivine, and ilmenite, and the estimated amount in successive residual liquids rose only slowly.

In the Palisades Sill, Ga as Ga^{3+} (0.62) in plagioclase probably entered the smaller Al^{3+} (0.50) sites because these elements are chemically similar, even though there is a large size discrepancy. That in the opaque iron minerals probably entered Fe^{3+} (0.64) sites. A coherence relationship has not been established, but the relationship is appropriate as Ga appears to form a stronger bond than Fe and their ionic sizes are close.

Tin and Lead

Tin. Tin is an element that appears relatively unimportant in terms of fractionation studies in the Palisades. It is present in most products of fractionation in concentrations bordering the detection limit of 1 ppm by the analytical method used (Table 13). Tin has not been determined in the various mineral phases. Preliminary analytical work suggests that it shows little or no variation with fractionation.

Lead. Lead distribution in the intrusion is irregular, and has been studied only in the whole rocks. The original magma contained only 11 ppm, and Pb shows a slight general increase in concentration with fractionation (Fig. 37). The average composition of the intrusion is 18 ppm. Values are given in Table 13. Lead tended to concentrate in residual melts, though some of it apparently entered crystal lattices and possibly occurs in K-feldspar and micas. Probably most of it is present as the sulfide phase. Galena has been observed in places in the upper levels of the Palisades intrusion in calcite veins.

PETROGENESIS

Dolerite sills provide an important insight into the processes operating in the petrogenesis of some basic magmas. The mineralogical and textural features reflect the physical conditions of rock formation, and depend largely on magma composition and its depth of crystallization. Because dolerites crystallize over a small temperature interval in shallow environments where physical conditions of crystallization change rapidly, many reactions between liquid and solid phases are arrested in intermediate stages. The plutonic intrusions of comparable composition provide many of the final or equilibrium products of these reactions, whereas the volcanic equivalents and, to some extent, the chilled contact dolerites reveal detail on early stage and metastable mineral relationships, and on the pre-emplacement history of the magma. The interpretation of the petrogenesis of the Palisades Sill relies only in part on measurement, and remains a subjective assessment of the data. However, it is hoped that the foundations have been laid for additional work, and that further measurement will, in due course, provide a more complete understanding of its petrogenesis.

An Evaluation of Gravitation Differentiation as Originally Applied to the Palisades Sill

Walker (1940) agreed with Lewis (1907, 1908a), and with many other geologists who have made general observations on the petrogenesis of the Palisades Sill, that crystal settling under gravity was the dominant factor controlling differentiation, and that differentiation was assisted by the early abstraction of olivine and by the relative upward migration of the liquid phase. This hypothesis found favor mainly as an explanation for the formation of the Mg-olivine layer. But for reasons given below, it is this feature of the intrusion which seems to be unsatisfactorily explained by a simple gravitation hypothesis. Indeed, it appears that some other events also contributed to its formation, as numerous intrusions of comparable size and similar composition are still preserved in the Palisadan province, but only the Palisades Sill contains a distinct layer, and then only in part of the intrusion.

The arguments advanced by Walker (1940, 1952, and 1956) for gravitational settling being the main mechanism in the differentiation process appear inconsistent with the time-space relationships between mineral phases observed by him. He (1940, p. 1085) has argued, on the one hand, that

> Separation of olivine [large grains] probably ceased in the metastable region in the upper parts of the sill, which had become relatively siliceous by the abstraction of the orthosilicate. It is thought that in the lower portion of the sill, which now contained a concentration of orthosilicate, separation of olivine continued into the labile region, when a shower of small crystals (richer in fayalite) was precipitated.

He considered that the intrusion comprised a single emplacement of slightly oversaturated magma in which olivine, the first phase to crystallize, settled to form the layer. One percent by volume of olivine in the intrusion would result in a layer considerably thicker than that observed, and its settling would cause little relative upward movement of the liquid phase. Concentration of orthosilicate (large grains) in a layer would in no way alter the composition of the enclosing liquid unless some of it was resorbed. Walker believed that some was. But he was under the impression that the small grains were more Fe-rich than the large. It seems unlikely that the temperature difference between the top and bottom halves of a simple dolerite intrusion immediately after emplacement would be sufficiently different to cause the settling olivine to resorb and then recrystallize again as olivine. It is more likely that such a magma would recrystallize pyroxene.

Walker has argued, on the other hand, that the sinking of olivine under gravity was unimpeded by the other mineral phases, because the crystallization of olivine was virtually complete before significant amounts of pyroxene and plagioclase crystallized. But he (1940, p. 1087) follows this discussion with the conflicting view that

> it seems likely that the crystallization range of the two minerals [olivine and pyroxene] overlapped to some extent, for there is a concentration of pyroxene in the lower part of the sill commencing considerably below the olivine diabase layer. Such a concentration could be caused only by the gravitational settling of pyroxene crystals to a growing lower chilled phase aided to a certain extent perhaps by reaction of liquid magma with sunken olivine crystals.

Thus it seems that he too was in considerable doubt about the timing and the sequence of events, for his observations are neither compatible with the observed paragenesis nor with the height of the layer, which, on the basis of his argument, should be below the main pyroxene concentration in the intrusion.

Furthermore, he submitted that the upward growth of the lower solidification surface accounted for the lack of packing of olivine, and emphasized that the margins of the layer are gradational. This is a matter of judgment, and I agree that small amounts of olivine are thinly distributed below the layer and, in much lesser amount, scattered grains occur for a few feet above the layer. But these distributions beyond the margins of the main layer are relatively insignificant when compared to the amount in the layer; the sharpness of the top margin is consistent with a sudden cessation of olivine crystallization.

Many of the observations made in the present investigation also proved inconsistent with previously held views. Most of the olivine grains in the layer appear too Fe-rich, too small and uncorroded, when considered relative to the height of the layer above the base, to attribute their concentration to settling after crystallization in a simple intrusion of oversaturated tholeiite. On an average they are much more Fe-rich than the corroded microphenocrysts in the basal chilled dolerite, and thus formed independently of them. However, similarities in the variation of the composition of the large and small grains in the layer, and their textural relationships with other minerals, are consistent with these belonging to the same generation of formation; their textural relations indicate that they are both of early crystallization, and measurement shows that, on an average, the large grains are slightly more Fe-rich than the small.

The height of the layer has always called for explanation, for many feel that had the original microphenocrysts of olivine, or grains of an initially crystallizing olivine phase, actually settled in a simple intrusion, the resulting layer would reasonably be expected nearer the base than it is. The experimental observations of Bowen (1915), and observations on natural phenomena (Richter and Moore, 1966), indicate that olivine in basic magma will sink fairly rapidly if unimpeded. Furthermore, under normal conditions of differentiation, including gravitational differentiation, in a simple intrusion equilibrium would have been established between mineral phases and liquid at the height the layer occurs; but as has been detailed, significant reversals occur in the fractionation trends of each mineral series in the layer, indicating a complicated history of crystallization for the intrusion. The discovery of an internal chilled contact at Haverstraw, whose features are consistent with those of the layer, points to another more probable explanation for the height of the layer in the intrusion; that is, the intrusion is a multiple one. The chemical evidence certainly supports this. In the case of a multiple intrusion, regardless of whether the olivine settled or formed *in situ* in the layer, the height of the layer would be determined by the location of the second phase relative to the first, because settling of olivine grains below the layer would have been inhibited by the advance of crystallization of the first phase up from the bottom contact; indeed, the advancing crystal framework was probably approaching the level at which the layer formed at the time that olivine crystallization of the layer commenced. The olivine grains strung out beneath the layer are in dolerite that contains some micropegmatite and pigeonite; their composition has not been measured yet, so that it is not known whether they are related to the olivine in the layer, or whether they crystallized from the first intrusive phase and settled. Other large intrusions in the province, such as the Lambertville Sill described by Jacobeen (*see* Hess, 1956, 1960) indicate that olivine concentrated only sparsely toward their base.

Walker (1940, p. 1083) has advanced numerous valid arguments to discount the possibility that the layer is a subsequent picritic intrusion into a virtually consolidated sill by explaining that although there are textural distinctions between the layer and the dolerites above and below, crystallization appears to have been continuous from the dolerite below to that above, and that the columnar joints pass uninterrupted through the layer.

The Internal Chilled Contact

At Haverstraw a reversal in the fractionation trend of the Ca-poor pyroxene series can be seen in the two diamond drill cores from holes 14 and 19 at 105 and 150 feet, respectively, from the base of the intrusion. The contact is marked by a pigeonite dolerite directly underlying a bronzite dolerite, each having merged so that crystallization appears to have been continuous between them. These contact dolerites have been described in the section on petrography and the contrast between them can be seen in Plates 9 and 10. Reaction rims did not develop on any of the minerals in them, but complicated crystallization is indicated in the bronzite dolerite by oscillatory zoning in some mineral grains, by the growth of augite and a few pigeonite crystals on the somewhat corroded bronzite microphenocrysts in a similar manner to that seen in the early dolerite at Englewood Cliff, and by the numerous patches of mesostasis, independent primary pigeonite crystals, and serpentine after olivine in both the dolerites. These features are absent from dolerites below and above the internal contact dolerites. That below shows progressive variation with fractionation from an early dolerite stage above the basal contact of the intrusion up to the early pigeonite dolerite stage at the internal contact, and that above again shows progressive variation with fractionation from the bronzite dolerite to an intermediate pigeonite dolerite stage at 432 feet, the limit of the diamond drill core examined.

Silicate analyses of the pigeonite dolerite (14-105) and the bronzite dolerite (14-109½) are presented in Table 8, where it can be seen that there is little difference in their major elemental composition. The trace element variations in the diamond drill core from the basal contact to 432 feet above

TABLE 17. ANALYSES OF ROCKS, HAVERSTRAW QUARRY

Specimen	Height*	Fe (%)	Mg (%)	Ca (%)	Mn (%)	Ti (%)	Ba (ppm)	Co (ppm)	Cr (ppm)	Cu (ppm)	La (ppm)	Ni (ppm)	Sc (ppm)	Sr (ppm)	V (ppm)	Y (ppm)	Zr (ppm)
14-5	5	7.6(s)	4.6(s)	6.8(s)	0.125(s)	0.63(s)	158	55	365	102	22	113	36	145	210	28	115
14-20	20	8.5(s)	5.3(s)	7.9(s)	0.13(s)	0.67(s)	158	60	370	110	26	143	38	160	215	32	125
14-50½	49	7.8(s)	4.6(s)	8.4(s)	0.13(s)	0.66(s)	150	52	380	110	35	96	39	170	220	36	125
14-80	78	7.8(s)	4.6(s)	7.6(s)	0.12(s)	0.67(s)	152	51	415	106	22	115	36	160	210	29	125
14-100	98	7.8(s)	4.9(s)	7.7(s)	0.13(s)	0.67(s)	178	57	440	128	28	130	38	160	215	32	125
14-104	102	7.84(c)	4.85(c)	7.36(c)	0.11(c)	0.62(s)	165	56	450	123	22	123	37	155	210	28	125
14-112	110	7.97(c)	4.58(c)	7.31(c)	0.12(c)	0.69(s)	185	56	345	123	17	113	36	155	240	26	130
14-120	118	7.8(s)	5.0(s)	7.3(s)	0.125(s)	0.63(s)	163	57	450	82	20	135	37	150	210	27	115
14-140	137	7.3(s)	4.9(s)	6.6(s)	0.11(s)	0.58(s)	140	60	425	88	25	155	35	140	190	29	100
14-200	196	7.8(s)	3.7(s)	7.5(s)	0.115(s)	0.61(s)	172	48	265	97	29	83	35	165	200	32	125
19-440	432	9.4(s)	3.1(s)	7.3(s)	0.13(s)	0.75(s)	225	47	41	138	39	55	34	185	210	36	160

(c) Chemical determination
(s) Optical spectrograph determination
* Height above base in feet

were also measured, including in detail the section across the internal contact. Values are given in Table 17, and plots of distribution trends are shown in Figure 38.

Interpretation of the trace element variations across the contact is complicated by the mineralogical differences between the dolerites, for it has been shown that the mineralogy of a rock influences the distribution of some elements. However, it does appear that other factors also influenced the variations shown. The trace elements least sensitive to mineralogical changes in the early fractionation stages—Y, La, Ba, V, and Cu—show significant inflections in their trends, which probably indicate a reversal in their fractionation trends. Thus these trace elements define the contact between the two

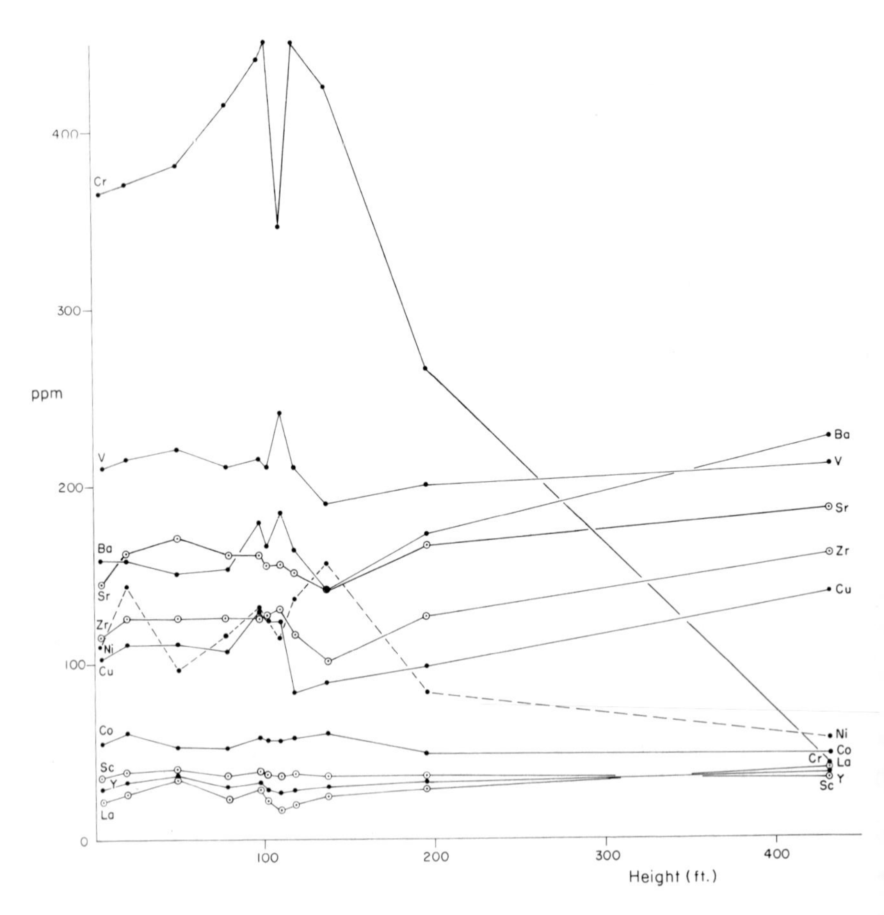

Figure 38. Distribution trends of trace elements in core of d.d.h.14, Haverstraw Quarry. Element values shown at 432 feet are of core from d.d.h.19.

magma phases and indicate some of the differences in the composition of the dolerites at the contact.

There is a possibility that the internal chilled contact is also present at Kings Bluff, though further field work and specimen collection would be necessary to confirm this. Here, a variant of the basal chilled dolerite occurs about 25 feet above the base of the intrusion, and the gabbroic olivine layer at 35 feet above the base rests directly on it. This chilled phase is distinct from the basal chilled dolerite in that it contains small plagioclase laths, which show a rough flow orientation, and glomeroporphyritic aggregates of euhedral plagioclase microphenocrysts; these are additional to microphenocrysts of olivine and glomeroporphyritic augite. However, the olivine is more abundant and less altered than that in the basal chilled dolerite, and the augite microphenocrysts commonly have nibbled grain margins owing to resorption; the orthopyroxene is of late crystallization—it occurs as reaction rims around some of the olivine grains, and also as large incomplete grains with numerous inclusions of groundmass grains.

Magma History before Emplacement

The main guide to the pre-emplacement history of the magma is provided by the microphenocrysts in the basal chilled dolerite, but this relates only to the original magma phase. Examination of this contact dolerite at six points between Kings Bluff and Haverstraw shows that it contains olivine, pyroxene, and rare plagioclase microphenocrysts. Those of olivine are either euhedral, skeletal, or resorbed, and at Englewood Cliff they have an average composition of Fo_{80}. Most of the pyroxene microphenocrysts are augite with slightly resorbed grain margins, though a few are euhedral or subhedral. They commonly form glomeroporphyritic aggregates that may also include an occasional bronzite grain. Otherwise orthopyroxene forms relict cores in a few of the augite grains. Because of the uniform composition and mineralogy of the chilled dolerite, and the consistent distribution of microphenocrysts, Hess (1956) has reasonably concluded that the magma emplaced was homogeneous and rapidly injected.

Microphenocrysts in the chilled dolerite presumably represent the solid phases in equilibrium with the liquid immediately prior to the emplacement of the magma. During emplacement many suffered mechanical deformation and strain. Texture, grainsize, and metastable mineral associations indicate that the contact dolerite chilled rapidly. Olivine, in particular, and to a limited extent augite, showed some instability in the sill environment, but in the case of pyroxene this is more noticeable in grains a small distance from the contact (25 to 30 feet) where cooling was a little slower, and where the number of pyroxene aggregates increased through settling. A few of these, including individual bronzite grains, contain an outer zone of freshly crystallized augite; others are enveloped by small augite and a few pigeonite grains. The impression conveyed by the habit of the microphenocrysts in the marginal zone of the sill is that the sill environment was not greatly different from the environment where the magma was located immediately before its emplacement. Probably temperature was the most significant difference, and variation in olivine composition, in particular, is sensitive to temperature change.

Hess (1956, 1960) has estimated that dolerite magma must lose about 300°C between its point of generation at depth, where a temperature of about 1400°C would be expected, and its emplacement in a sill, because the indications are that in a sill crystallization commences at about 1100°C.[6] Assuming this temperature for the Palisades, he estimates that the magma would have contained less than 50°C superheat because sedimentary xenoliths in it are neither melted nor assimilated. Thus, because tholeiitic magma appears to move rapidly during igneous activity, the magma that formed the sills of the Palisadan province, including the flows of the Watchung basaltic succession, must have paused at some stage in transit. Presumably mobilized magma mush was repeatedly forced up from the point of generation at depth into a fairly shallow reservoir, where it became mixed and lingered sufficiently long to establish a fractionation gradient and equilibrium between the liquid and solid phases. During the pause in the reservoir the magma lost much heat by conduction through its walls and by the melting of crystals incorporated at depth, both of which are heat dissipation mechanisms suggested by Hess (1956). The intermediate reservoir developed in an environment where olivine of composition Fo_{80} crystallized in equilibrium with a tholeiitic liquid containing about 3 percent normative quartz. Magma of this fractionation stage was expelled first. The reservoir expelled magma of a more advanced fractionation stage containing about 5 percent normative quartz when the second phase was emplaced.

These observations are consistent with those of Richter and Moore (1966) and Murata and Richter (1966) on the Kilauean province where igneous activity is manifested in a series of eruptions and injections, and where magma appears to be tapped at different fractionation stages from a shallow reservoir. In this case, however, the more acid phase erupted first. They also observed that the extruded magma recycles with reservoir magma between eruptive phases, and that crystals in the erupted magma tend to react with the liquid in the surface environment.

---

[6]It would be difficult at present to hazard an estimate of the temperature of the magmas emplaced in the Palisades Sill by the procedure used by Hess (1960, Fig. 12, p. 40 and 78) for plutonic intrusions of comparable composition, because the crystallization of the pyroxenes was complicated by multiple intrusion in the Palisades.

The Formation of the Mg-Olivine Layer and the Internal Chilled Contact, and the Petrogenesis of the Sill

Many of the clues on the petrogenesis of the intrusion are in the Mg-olivine layer and the internal chilled contact. The following discussion suggests the magmatic conditions and factors which may have led to their formation, but both are still under investigation (Walker and Green, in prep.). The discussion of their formation also gives an insight into the petrogenesis of the sill.

The layer and the internal contact have been described in detail, and it has been shown that the occluded chilled contact in the Haverstraw area has some similar features to the Mg-olivine layer; a common origin for them seems probable. It is believed that both formed at the junction between two magma phases, the first being partly crystallized and differentiated before the second was emplaced. It was during the crystallization which followed that the Mg-olivine layer formed, and the reversals developed in the fractionation trend of each mineral series, both in the layer and at the internal contact. When equilibrium was established between the phases, differentiation proceeded on a normal course toward iron-, silica-, and alkali-enrichment.

Before presenting the possible reasons for the formation of the layer and the internal contact, it should be appreciated that there may have been some factors contributing to their formation whose effect it would be virtually impossible to assess at present. For example, the changing form of the intrusion northward may have contributed to the variation in the contact relationship between the two magma phases. It should be recalled that the intrusion is sill-like between Jersey City and Nyack Beach State Park, and that the Mg-olivine layer occurs from 25 to 60 feet above the basal contact, but that it is somewhat variable in height and thickness, and it tends to become higher, and beyond Forest View Lookout, thinner, toward the north. In the dike-like part of the intrusion, where the Mg-olivine layer is absent, the internal contact is between 105 and 150 feet from the basal contact where observed at Haverstraw. It is possible, therefore, if the second phase of magma entered the intrusion in the southern part, that the injection focused some of the Fe-rich and cooler fraction of the first phase into the dike-like part of the intrusion.

Another factor that is impossible to assess is the effect of tectonism on the accompanying igneous activity, and its influence on the movement of magma during emplacement and consolidation. In addition, there could have been changes in pressure and escape of volatiles through fissures, and, if these were of sufficient magnitude, they could have modified the course of crystallization in the intrusion. A sudden release of volatiles could cause an increase in magma viscosity, for example, and lead to undercooling, and to supersaturation in a particular mineral phase (and there may have been a tendency for this effect to occur at the contact between the two magma phases), whereas an abundance of volatiles in the magma could cause a variation in the stability field of a mineral phase by depressing its liquidus in an equilibrium assemblage. Possibilities such as these cannot be evaluated at present, though textural features may be a guide to their interpretation when more detailed measurement has been made on mineral compositions.

Excluding the late dikes, chemical evidence shows that the intrusion contains two magma phases. The second was much larger and of a later fractionation stage than the first, and it appears to have intersected the first above the chilled dolerite stage at Kings Bluff, the early dolerite stage at Englewood Cliff, and an early pigeonite dolerite stage at Haverstraw. On the grounds of mechanical efficiency it is unlikely that the second magma phase sandwiched laterally into a hot but completely crystallized first phase, wedging it apart along its length; it is more probable that the magma gained access by exploiting the weakest and mainly liquid portion of the original intrusion.

As mentioned, it is believed that the Mg-olivine layer and the internal chilled contact are features which developed between the two main magma phases during largely continuous crystallization of the intrusion. The contact between the phases in the sill-like part of the intrusion was probably slightly below the layer, whereas it is clearly marked in the dike-like part by a sharp reversal in the trend of the pyroxene crystallization. The difference in the expression of the contact probably depended largely on the composition of each phase and its variation in temperature along the length of the intrusion upon emplacement. These would probably be the main controlling factors on whether or not olivine crystallized, and they will be elaborated on below.

The simplest contact relationship occurs in the dike-like part of the intrusion at Haverstraw. Here, the contact reveals most about the second magma phase. Pseudomorphs after olivine in the dolerites immediately below and above the contact suggest that the second magma contained microphenocrysts upon emplacement, and the altered glassy mesostasis suggests that some undercooling occurred. Thus it seems reasonable to assume that the second phase, like the first, was rapidly emplaced and probably homogeneous. Possibly the bronzite microphenocrysts in the dolerite above the contact were introduced with the magma, even though they merge with the orthopyroxene crystallizing from the liquid in the intrusion within 25 feet above the contact. This would imply that they were largely in equilibrium with the magma, except at the contact where augite and some pigeonite molded on their somewhat corroded margins, in a similar manner to that shown by the bronzite microphenocrysts in the early dolerite of the first phase at Englewood Cliff. Crystallization in the second phase at Haverstraw probably commenced with plagioclase and monoclinic pyroxene, though orthopyroxene followed closely. As mentioned, crystallization in the first phase at the contact was that of an early pigeonite dolerite stage; slight mixing at the contact is indicated by some pigeonite grains of the crystallizing first phase being included in the bronzite dolerite.

If pressure can be assumed to be constant immediately after the injection of the second phase, the mineralogical differences between the dolerites below and above the contact result mainly from the difference in their temperature of crystallization because analyses show that the major elemental composition of both are very similar (Table 8). According to the plot relating the pigeonite-orthopyroxene inversion curve to composition and magma temperature (Hess, 1960, Fig. 12), about a 25°C rise in temperature in the contact zone would have been sufficient to move from the pigeonite to a bronzite stability field with progressive crystallization.

If the apparent internal chilled dolerite at Kings Bluff is that of the second phase, more information on this phase is revealed. It contains more plagioclase microphenocrysts and orthopyroxene than the basal chilled dolerite of the first phase. This suggests that the second magma phase was cooler than the first upon emplacement, for Richter and Murata (1966) have shown that the coolest and most Si-rich Kilauean lavas were those with plagioclase and orthopyroxene microphenocrysts in addition to olivine. Cooler emplacement is also consistent with the second magma being of a more advanced fractionation stage than the first and with the bronzite microphenocrysts (En_{78}), above the internal contact at Haverstraw, being introduced with the second phase magma, for their composition is more Fe-rich than those (En_{83}), in the early dolerite of the first phase above the chilled selvage at Englewood Cliff.

The most complex relationships between the magma phases are shown in that part of the intrusion where the Mg-olivine layer consistently developed above the early dolerite stage of the partly crystallized first phase. In the early dolerite of the first phase, there is a slight concentration of microphenocrysts of olivine and glomeroporphyritic pyroxene above the basal chilled selvage, and many of these apparently settled from the overlying liquid. Their partly resorbed condition suggests that although they were not entirely in equilibrium with the crystallizing liquid, it is unlikely that many were completely resorbed. Augite and some pigeonite built on the partly resorbed bronzite microphenocrysts, and, as mentioned, their habit is similar to those in the dolerite above the internal contact at Haverstraw. The indications are at Englewood Cliff that crystallization in the first and more basic phase commenced with monoclinic pyroxenes and plagioclase, and that these were followed by orthopyroxene. Apparently the temperature of this phase upon emplacement in the Englewood Cliff area was not high enough for olivine to crystallize; that which settled was introduced by the first magma phase.

Higher in the intrusion at this locality the development of the Mg-olivine layer in the zone of the contact between the two magma phases resulted in more complex mineral relationships than elsewhere. The layer is texturally distinct from the dolerites below and above. The olivine grains in the layer appear to belong to a single generation of formation. They are uncompacted, unlaminated, and do not show grading according to size. Dielectric anisotropy tests (F. Stacey, 1961, written commun.) show that the olivine grains in the layer are isotropic, and thus have no particular orientation. The large grains are in greatest concentration at the base of the layer, and their habit gives the impression that they have settled. The numerous small grains are mainly in the plagioclase areas of crystallization and are greatest in concentration, and richest in the forsterite molecule in the center of the layer. Where they are within a single large normal zoned plagioclase crystal, they individually show normal zoning, and are progressively more Fe-rich from the center to the margin of the plagioclase. These features together with the systematic variations through the layer shown by all the mineralogical, compositional, textural, and zonal features, as previously detailed, indicate that crystallization of the layer was largely continuous with that of the dolerites below and above. Crystallization in the layer commenced with olivine, early plagioclase, and augite, and these were followed by orthopyroxene.

Observations made so far on the Mg-olivine layer are consistent with the following hypothesis for its formation. In the sill-like part of the intrusion much of the first magma phase was in an early stage of fractionation when the second was emplaced. Following the introduction of the second phase, a cooling gradient developed across the contact between the phases because of the differential in their temperatures. Similarly another, and probably more severe, gradient developed across the top contact of the second phase. In addition to these contact gradients, there was a significant thermal gradient across the second intrusion, which suggests that it was emplaced by a series of injections which directed the hottest magma into the top half. Because of the gradient, the initial olivine crystallizing toward the base was in the range Fo_{63} to Fo_{55}, whereas that crystallizing in the top half was in the range Fo_{77} to Fo_{69}. In the late stages of olivine crystallization, the top half yielded olivine in the range Fo_{69} to Fo_{58}.

Thus olivine of contemporaneous crystallization varied consistently in composition at successively higher levels in the second intrusion according to the prevailing thermal gradient across the intrusion. If most of the olivine that crystallized has been preserved in the Mg-olivine layer, the progressive variation in olivine composition from bottom to top of the layer probably crudely reflects in miniature the variation in the thermal gradient through the second phase intrusion during the period of olivine crystallization. It would be very difficult to determine, however, the form of the gradient from this variation in olivine composition because of the difficulty in estimating the time of fall of olivine to successive levels in the layer.

Olivine crystallization in the second phase magma was immediately accompanied by early plagioclase and probably early augite crystallization. These phases grew by progressive zoning and compounded together during their fall to the advancing framework of crystallization at the bottom cooling surface of the sill, where they lodged. As mentioned, the successive arrivals were in an order consistent with the compositional variations observed in the layer. The layer would have thickened rapidly relative to the normal advance of crystallization in the intrusion by fractional crystallization because it is the residual accumulation of all the heavier-than-liquid phases that crystallized during the period of olivine crystallization. With the cessation of olivine crystallization the layer was abruptly completed, and the main stage of gravitational differentiation in the intrusion tapered off.

After the rapid build-up of the layer progressive crystallization would continue in the layer until all the interstitial liquid had crystallized. During this period, plagioclase and augite crystals would continue to grow with the additions of zones of more advanced crystallization. With the enlargement of these grains much of the orthopyroxene crystallized, and in the residual stages biotite formed (Pl. 5). Although the formation of the layer is visualized as being similar to that of an orthocumulate (Wager and others, 1960), it differed in that some exchange would probably have taken place between the interstitial liquid and the overlying magma because of the limited thickness of the layer.

The layer accumulated during the quiescent magma state that existed while the temperature variation in the multiple intrusion was establishing a cooling gradient consistent with that which normally exists in a simple sheet intrusion during crystallization. Thus, while thermal equilibrium was being established between the magma phases, a pause occurred in the general cooling trend of the intrusion, during which time the layer and its distinctive gabbroic texture developed. After the formation of the layer progressive crystallization of the intrusion continued, and fractional crystallization proceeded on a normal course toward iron-, silica-, and alkali-enrichment until crystallization was complete.

All current observations on the intrusion appear to be satisfied by the above hypothesis, except possibly the co-existence of pigeonite with orthopyroxene in the dolerite immediately above the layer. An understanding of this association requires more measurement on the mineral phases involved to establish whether or not disequilibrium on metastable conditions during crystallization are indicated. If either were the case, it could be argued that equilibrium had not been reached between the magma phases by this stage in crystallization, and possibly that a flow differentiation origin for the layer should be considered. However, this explanation, in turn, is difficult to reconcile with the other evidence, in particular, with the systematic mineralogical and compositional variations through the layer. Possibly, therefore, the association indicates that some form of stable co-existence occurred between these minerals during this stage in crystallization, which would give some meaning to the configuration of the pyroxene fields shown in Figure 7.

The validity of the above hypothesis also depends in part on whether olivine can crystallize under natural conditions from magma with a composition of the second phase. Evidence from the Hawaiian province, and, in particular, from the 1959-1960 summit and flank eruptions of tholeiite at Kilauea (Murata and Richter, 1966, and Richter and Murata, 1966) would suggest that the initial crystallization of olivine is feasible in magma with compositions similar to those in the Palisades Sill. Table 18 compares the average composition of the original magma and that of the Palisades intrusion with two of the most fractionated flank lavas of the Kilauean eruptions, and indicates that both magma phases in the Palisades are similar in chemical and phenocrystic composition to the Kilauean lavas, except that they contain slightly more Si and less Ti. The Kilauean lavas have up to 2.7 percent normative quartz and contain olivine phenocrysts with compositions up to Fo_{76}. The liquid from which these phenocrysts crystallized has 3.8 percent normative quartz. Murata and Richter believe that tholeiitic lavas of the type quoted crystallize olivine as an initial phase if they have more than 6 percent MgO. In this respect dolerites of corresponding composition probably behave similarly, as olivine crystallization in shallow environments is mainly a function of magma composition and temperature. Pressure difference is probably too small to have a significant influence. Thus, it seems possible that olivine could have crystallized as the initial phase from a magma with up to 5 percent normative quartz.

As mentioned, the temperature of each magma phase after its emplacement largely determined whether or not olivine crystallized. In the sill-like part of the intrusion, where the second magma phase came in contact with the first it was capable of crystallizing olivine. However, in the dike-like part pyroxene was the initial phase to crystallize, which suggests that the temperature of the magma above the internal contact was lower than that of the magma above the Mg-olivine layer, for the chemical data indicate that there was not a significant difference in the composition of the second magma in the contact zone. Thus, the coolest portion of the second magma entered the northern part of the intrusion.

TABLE 18. COMPARISON IN COMPOSITION BETWEEN PALISADES AND KILAUEAN THOLEIITES

		1	2	3	4	5
	SiO_2	52.0	52.35	50.58	50.62	50.69
	TiO_2	1.2	1.6	3.11	3.20	3.56
	Al_2O_3	14.5	14.0	13.88	13.76	13.51
	Fe_2O_3	1.35	2.65	2.18	1.80	1.87
	FeO	8.9	9.2	9.52	9.95	10.70
	MnO	0.15	0.15	0.18	0.18	0.18
	MgO	7.6	6.15	6.56	6.43	5.74
	CaO	10.3	9.35	10.43	10.21	9.72
	Na_2O	2.0	2.5	2.58	2.63	2.60
	K_2O	0.85	0.85	0.65	0.65	0.72
	P_2O_5	0.15	0.25	0.35	0.35	0.39
	H_2O	1.05	1.3	0.09	0.10	0.08
		100.05	100.35	100.11	99.88	99.76
				NORMS		
	Q	2.8	4.6	2.69	2.49	3.83
	Or	4.7	5.4	3.84	3.84	4.24
	Ab	16.9	21.0	21.43	21.98	21.84
	An	28.2	24.2	24.58	23.97	23.15
	Wo	9.3	8.7	10.31	10.11	9.31
Di	En	5.2	4.7	5.91	5.59	4.83
	Fs	3.7	3.7	3.94	4.13	4.25
Hy	Fs	9.85	8.7	6.94	7.70	8.31
	Mt	2.0	3.8	3.16	2.61	2.71
	En	13.75	10.6	10.42	10.42	9.46
	Il	2.3	3.0	5.91	6.08	6.76
	Ap	0.25	0.5	0.83	0.83	0.92
	H_2O	1.05	1.3	0.09	0.10	0.08
		100.0	100.2	100.05	99.85	99.70

MICROMETRIC ANALYSES (Microphenocrysts only in Vol. %)

	1	3
Ol	1.5(Fo_{80})	2(Fo_{79})
Cpx	1.0	6
Opx	0.5($En_{83?}$)	1(En_{80-75})
Pl	0.1(An_{66})	9(An_{72-66})

(1) Average Palisades basal chilled dolerite

(2) Average composition of the Palisades intrusion, composite section, a planimetric estimate from 0 to 1000 feet, Figure 10

(3) 1960 Kilauea early flank eruption—slab pahoehoe from west vent (Murata and Richter, 1966, Table 2, Anal. F-1). Modal composition from Richter and Murata (1966, Table 1, Anal. F-1)

(4) 1960 Kilauea early flank eruption—pumice from main vent (Murata and Richter, 1966, Table 2, Anal. F-3)

(5) 1960 Kilauea early flank eroption—basaltic glass of pumice Anal. 4. (Murata and Richter, 1966, Table 4, Anal. F-39)

The Differentiation Process

It has been established that the Palisades Sill is a multiple intrusion comprising at least two magma phases, into which late-stage dikes intruded after the main phases had consolidated. This seems reasonable, as the contemporaneous Watchung basalt flows, with three main basaltic successions, show that igneous activity at the time was protracted and comprised a number of phases.

Apart from the obvious influence of temperature, pressure, and magma composition on the factors controlling the differentiation process, most explanations of the process in dolerite sills to date asume that mechanical factors exert the main control, and much less attention has been paid to the chemical factors, probably because these are less well understood. Ignoring them, however, has led to inconclusive results from many differentiation studies. The mechanical factors most commonly accepted as important are the settling of solid phases under gravity and the upward displacement of the liquid phase, the movement of phases by gas streaming, convection, magma flow, and filter pressing, whereas the chemical factors advanced as important are fractional crystallization, diffusion, and immiscible liquid separation. Other important controls are more specific; for example, the partial pressure of oxygen (Osborne, 1959) and its influence on the trend followed in differentiation—Fe-enrichment or alkali-enrichment. The volatile and, in particular, the water content of the magma strongly influence viscosity and, in turn, nucleation and texture. In addition, water content may influence the early or late crystallization of some minerals; for example, the opaque iron minerals. It may also influence the mobility and distribution of elements in magma; for example, ferric iron is more water soluble than ferrous—and so on.

The factors mentioned are not exhaustive, but they serve to indicate that the differentiation process is a complex one of interacting factors. The problem is to ascertain which factors operated and the degree to which each contributed to the over-all differentiation of the intrusion. This depends, in part, on the duration of their operation, which, in turn, depends on the cooling rate of the intrusion. Some factors mentioned above apply only in certain situations; for example, flow differentiation (Simkin, 1967) and convection (Hess, 1960). Others are not clearly understood, which is the case for liquid fractionation, a term, introduced by Hamilton (1964, 1965) to describe the effects that he believes result from diffusion and liquid immiscibility in fractionation.

Some of the most important observations on differentiation in tholeiitic magma are those currently being made on the crystallization of the Kilauea Iki magma lake. The studies of Richter and Moore (1966) confirm a number of factors which geologists have previously believed important from their observations on rocks or from laboratory experiments. By studying the advance of the growing framework of crystals with the cooling of the magma lake, Richter and Moore have been able to follow progressively the differentiation process, and have observed in operation gravitational differentiation, filter pressing of interstitial liquid from the crystal framework into fractures, and the concentration of Si and alkalis toward the top of the liquid phase (which they suggest is aided by gaseous transfer). In addition, Murata and Richter (1966) examined the differentiation of the Kilauean lavas and showed that it is consistent with the concept of olivine control of magma composition (Powers, 1955), and with the separation of solid phases, in particular olivine, from the liquid during progressive crystallization.

The differentiation process proposed for an intrusion must be consistent with its texture and paragenesis. These are fairly complex in dolerites because cooling was sufficiently rapid to prevent equilibrium being established between liquid and solid phases during successive stages in fractionation; zoning in minerals bears witness to this. Thus dolerites preserve a detailed record of their cooling history in a series of exclusive textural and mineral relationships, all of which provide clues on their differentiation, but which must be understood if the differentiation process is to be defined.

It is believed that subophitic texture forms largely as a result of the rate of cooling of tholeiitic magma in sills. Though the major elemental composition mainly determines which minerals form, it is the volatile content in conjunction with changing temperature and composition of magma during cooling which largely influence viscosity in the liquid phase, and thus the rate that crystals nucleate and grow.

In the Palisades Sill, subophitic texture occurs throughout, except in the Mg-olivine layer, which is gabbroic. With progressive crystallization, however, the subophitic texture became steadily coarser, and in the late-stage rocks it tends toward gabbroic. This is consistent with increasing volatile and $(OH)^-$ content in the liquid and the consequent decrease in viscosity during progressive crystallization. The systematic variation in texture with fractionation, and the almost universal interlocking of plagioclase and pyroxene grains, imply that these minerals crystallized together and progressively with cooling, and that relative movement between them under gravity was unlikely during much of the crystallization. The movement of small aggregates of pyroxene and plagioclase crystals under gravity, proposed by McDougall (1962, 1964) as a modification of the roof-stoping and mush-settling hypothesis put forward by Jaeger and Joplin (1955), if it occurs, would be limited to the early fractionation stages because such movement in the later stages does not appear consistent with the textural features of dolerite sills.

In the Palisades the microphenocrysts of olivine and pyroxene, including the glomeroporphyritic aggregates, are the solid phases most noticeably affected by gravity settling, for in some cases they appear to be foreign to the dolerites in which they occur; for example, in the early dolerite. However, the Mg-olivine layer is largely homogeneous in appearance, though it is the accumulation of olivine enveloped in early plagioclase and augite. It appears that after its formation, gravitational differentiation tapered off, as textural evidence indicates that subsequent settling of heavier-than-liquid phases must have been limited to short-distance gravity settling. The relative upward displacement of the liquid phase as a result of the combined movement of solid phases under gravity was probably small in amount relative to the over-all differentiation.

Textural, paragenetic, and chemical evidence indicate that fractional crystallization was more important than gravity settling of solid phases in the differentiation of the sill because it operated throughout crystallization. Fractional crystallization was largely responsible for the gradational layering of the intrusion according to progressively lower temperature assemblages upward. The study of element behavior with progressive crystallization has indicated that fractional crystallization is a process of selective removal of elements from the liquid phase and their regular arrangement in the solid phases according to the systematic variation in their chemical properties. Elements not readily accepted in the early-forming minerals tended to migrate upward until suitable physicochemical conditions prevailed for their crystallization. The chemical factors controlling element removal from the liquid have been discussed in the section on trace elements, where it has been shown that they strongly influence element distributions during progressive crystallization. The distributions, however, are modified by other interacting factors in the over-all differentiation process; settling of solid phases, for example, exerts a significant modifying influence in the early fractionation stages, and gravity would also influence the distribution of the liquid phase during crystallization where convection or liquid immiscibility are factors in the differentiation process.

Crystallization of the sill proceeded most rapidly from the base of the intrusion, where the most refractory phases occur. At the same time crystallization advanced from the top contact, but at a slower rate, so that the advance of the frameworks of growing crystals met at about six-sevenths of the way through the intrusion from the bottom contact (Fig. 6). Jaeger (1957) has estimated that cooling of an intrusive sheet takes place from both contacts at approximaely the same rate in the absence of convection, provided that the thickness of cover is greater than half the thickness of the sheet. Cover on the Palisades Sill would have been well above this limit, but progressive crystallization from the contacts was grossly asymmetric. The settling of solid phases contributed in a small way to this variation, but it is clear that some other factors in the differentiation process must be largely responsible; these are probably the ones that kept the liquid chemically active, and thus enabled differentiation to take place systematically by fractional crystallization during the progressive crystallization of the intrusion.

The chemical reactivation of the liquid phase in the vicinity of crystallization may occur partly in the way Hess (1956, 1960) has suggested. In dolerite sills it appears that differentiation takes place primarily by crystallization inward from the surfaces of cooling, with diffusion exchange between the interstitial residual liquid in the framework of growing crystals and the main mass of overlying liquid. The diffusion factor is important, for it makes possible the reactions that take place between the liquid and solid phases, and thus with the continually changing physico-chemical conditions during cooling, it enables the solid phases that progressively form to be preserved by protective zoning or more advanced crystallization. Chemical reactivation is probably also assisted by other factors. Where convection operates it would cause magma mixing and would tend to direct heat toward the top and coolest part of the intrusion, and concentrate residual liquid there. Gas streaming would also aid magma mixing and similarly contribute to the upward migration of residual material from crystallization. In addition, supercritical water, where present, would carry soluble phases in the direction of decreasing pressure in the intrusion. Richter and Moore (1966) have observed that Si and volatiles including alkalis are concentrating toward the top of the liquid phase in the Kilauea Iki magma lake. The residual products of fractional crystallization, together with the components in the liquid not participating in early crystallization, make the main contribution to these late-stage concentrations of Si and alkalis, and their upward migration appears to have been a common feature in the solidification of many dolerite sills. This feature of differentiation, for which Hamilton (1965) invoked liquid fractionation mainly to explain, can be considered reasonably within the scope of fractional crystallization, for the behavior of the liquid and solid phases during progressive crystallization are not independent, except in the case of liquid immiscibility, but are parts of the one chemical system. Element partitioning between phases goes according to the thermodynamics of the crystallization reactions in the magmatic system. The least refractory components remain largely in the liquid phase until suitable physico-chemical conditions prevail for their crystallization, and these occur mainly in the coolest or late stages of crystallization. Liquid separated from the main fractionation trend by immiscibility would be expected to follow an independent line of differentiation. The distinct change in the differentiation trend from iron- to alkali-enrichment during late-stage fractionation in dolerite sills (Fig. 11) does not, however, necessarily indicate two independent trends as a result of liquid immiscibility; it still appears to be logically explained in terms of fractional crystallization, and is probably in response to changes in other factors in the magmatic system, such as partial pressure of oxygen (Osborne, 1959).

Secondary factors operated during the consolidation which made minor contributions to the over-all differentiation of the intrusion. These resulted in the isolation of the pegmatitic schlieren and the late soda-rich hydrothermal veins from the main fractionation sequence. The schlieren apparently represent pockets of liquid charged with volatiles which became isolated from the main fractionation trend, possibly as pockets of liquid immiscibility. Their independent crystallization is indicated by their chemistry. The major and trace element contents of a typical schliere are given in Table 13, and the differentiation trend of the schlieren relative to the main trend of the intrusion is shown in Figure 11. The soda-rich hydrothermal veins probably formed as a result of filter-pressing of late residual liquid into fine fissures that developed during the late consolidation stages of the intrusion. Their location in relation to the main fractionation trend is also shown in Figure 11. Late hydrothermal veins and alteration also indicate that hydrothermal liquids migrated where access could be gained during late consolidation.

SUMMARY AND CONCLUSIONS

The detailed field, mineralogical, petrological, and chemical study of the Palisades Sill has provided solutions to some of the existing problems of its petrogenesis and, in particular, of its differentiation. The additional study of element partitioning between mineral phases with progressive crystallization has yielded much new information on element behavior with fractionation in tholeiitic magma.

The Palisades Sill is about 1000 feet thick and 50 miles in length, and was emplaced during the Mesozoic in the Newark Formation of New Jersey and New York states; it is the intrusive equivalent of the Upper Triassic Watchung flows, and is transgressed by late chilled dolerite dikes. Its outcrop in New Jersey indicates that it is largely sill-like in form, but the outcrop in New York State shows that it also has some dike-like features.

The uniform mineralogy and composition of the chilled margins indicate that the original magma emplaced was homogeneous and rapidly injected. The impression conveyed by the habit of the olivine, augite, and rare bronzite and plagioclase microphenocrysts in the marginal zone is that the sill environment was not greatly different from the environment where the magma was located immediately before its emplacement; probably temperature was the most significant difference.

The rocks in the sill which represent the various stages in the fractionation sequence have been described and distinguished on the basis of petrographic type; they are all variants of tholeiitic dolerite and have been grouped according to early, middle, and late fractionation stages. Chemical evidence shows that the intrusion comprises two magma phases of oversaturated tholeiite. The second phase was much larger and of a more advanced fractionation stage than the first, and was emplaced above the crystallized part of the original phase. In both phases the order in which the essential minerals began to crystallize was olivine, plagioclase, augite, and orthopyroxene.

Away from the basal chilled dolerite crystallization with fractionation proceeded normally, except for the complex mineral relations that developed in the bottom quarter of the intrusion as a result of the two-phase intrusion of magma. The normal course of fractionation intiated by the first phase magma was interrupted by the introduction of the second, which caused reversals in the fractionation trends of most mineral series at the junction between the phases. Following adjustment to the changed physio-chemical conditions of crystallization, fractional crystallization again proceeded normally in the intrusion.

During the early and middle fractionation stages, the Ca-rich pyroxene crystallized in cotectic equilibrium with the Ca-poor varieties, and in late stages, beyond the so-called two-pyroxene field, two ferroaugites crystallized, one mauve-brown and the other pale green. Petrographic evidence shows that the mauve-brown ferroaugite is continuous with the normal augite trend, and that the pigeonite trend appears to give way to pale-green ferroaugite in early late-stage fractionation. Ferroaugite relationships are complicated, however, for both ferroaugites co-exist, even though some of the mauve-brown variety was involved in the pyroxene transformation to the pale-green variety. Moreover, some of the pale-green variety exsolved(?) Fe-rich orthopyroxene, and with progressive fractionation small grains of this phase possibly reacted with liquid to form fayalite and quartz in the very late stages. Green-brown hornblende most commonly developed in the place of the pale-green ferroaugite. These relationships, which are tentatively established on petrographic evidence, require more rigorous chemical investigation. The end products of the reactions appear to be largely determined by the hydrous state of the magma during crystallization. Possibly the pyroxene reactions in late-stage fractionation result in a bifurcation of the orthopyroxene-pigeonite trend to a pale-green ferroaugite trend and an orthopyroxene fayalite trend. The compositional change of both the Ca-rich and the Ca-poor pyroxenes with fractionation is to successive enrichment in Fe^{2+} relative to Mg, and to some variation in the Ca content.

Most of the orthopyroxene crystallized in the lowest quarter of the intursion, where two varieties are present. The first, the Bushveld type, crystallized directly from the magma and has a reaction relationship with early olivine. Its composition ranges from En_{83} to En_{60}. The second, the Palisades type, was derived by inversion of pigeonite, and ranges in composition from En_{70} to En_{50}, but is mostly between En_{60} and En_{50}. Thus, En_{50} was the composition reached before pigeonite remained the stable phase through cooling. The pale-green ferroaugite that followed pigeonite crystallization in the fractionation sequence exsolved(?) orthopyroxene whose composition ranges from $En_{35?}$ to $En_{25?}$.

Olivine occurs in two horizons of the fractionation sequence. That in the Mg-olivine layer is a concentration of the initial phase to crystallize from the second magma phase, and ranges in composition from Fo_{77} to Fo_{55}. That in the Fe-olivine horizon formed in the very late fractionation stages and possibly was derived, in part, by reaction from the Fe-rich orthopyroxene; it has a composition range of $Fo_{20?}$ to Fo_7.

Plagioclase in the dolerites ranges in average composition from $An_{63.5}$ to An_{37}; it became progressively enriched in the albite molecule with fractionation. Micropegmatite, including alkali feldspar, increased progressively in abundance toward the top of the fractionation sequence. Plagioclase crystallization, like that of augite, was continuous from early to late crystallization, and both are characteristically in subophitic relationship; this is the dominant texture of the intrusion. Other minerals, most of which also increase in abundance toward the top of the fractionation sequence, are hornblende, biotite, opaque iron minerals, apatite, zircon, chalcopyrite, and sphene. Movement of volatiles caused local alteration to minerals at all stages in fractionation.

The junction between the two magma phases is marked by the hyalosiderite layer, and in the northern part of the intrusion by an internal chilled contact where a reversal in the pyroxene fractionation trend occurs. The temperature and composition of the magma phases determined the amount of olivine that crystallized, and hence the expression of the contact along its length. It was only in the second phase in the southern part of the intrusion that sufficient olivine crystallized to form a distinct layer in the contact zone between the two phases. The olivine during its fall and accumulation was enveloped by early plagioclase and augite, and formed an orthocumulate at the base of the second magma phase. The layer has a distinctive gabbroic texture which probably indicates that it formed during a pause in the cooling trend of the intrusion. After equilibrium was reached between the magma phases, which was early in the fractionation sequence, fractional crystallization proceeded on a normal course towards iron-, silica-, and alkali-enrichment. The formation of fayalite granophyre and granophyric dolerite in the late products indicates that fractionation reached a fairly advanced stage for the magma type. The fractionation process toward more acid types was halted at the ferrodolerite stage by the advance of crystallization from the top contact. Comparison with the Karroo, Tasmanian, and Antarctic provinces shows that the composition, and the trend and range of differentiation in the Palisades, are very similar.

The differentiation in the Palisades Sill resulted from a combination of interacting factors that operated during progressive crystallization. The duration of their operation, which was largely controlled by the rate of cooling of the intrusion, determined the extent of their contribution to the over-all differentiation. Of the main factors involved, fractional crystallization operated throughout crystallization, whereas gravity settling of solid phases was mainly limited to the early fractionation stages. Additional control on fractional crystallization was the rate of diffusion exchange between the liquid and solid phases, and this was determined largely by the extent to which the liquid remained chemically active by mixing in the vicinity of crystallization. No direct evidence has been found in the textural features to determine how mixing was achieved; presumably it was a function of mechanical factors such as gas streaming and convection.

Trace element partitioning between mineral phases has been studied over a wide compositional range in each mineral series, and the element distributions and coherence relationships reveal how elements behave in a tholeiitic intrusion during fractionation. In particular, the geochemistry of Sr, Ba, Cr, V, Sc, Co, Ni, Y, Zr, La, and Cu has been studied in detail. Their behavior is determined by their properties in the magmatic environment of liquid and solid phases, where the ions, including those of the major elements, are competing for an ordered arrangement during the solidification of the intrusion. It is not only the properties of the ion in the liquid, but also its properties within the crystal lattice, particularly its bonding characteristics, which determine its behavior. Trace cations, like the major cations, show varying site preference according to their bonding characteristics in different lattice structures; for example, Co mainly follows Fe^{2+} in clinopyroxene and orthopyroxene, and Mg in olivine. Thus, it was found that the elements in a crystallizing magma will, so far as possible, distribute and arrange themselves in crystal lattices according to their total chemical properties, and that, *in general terms,* the cation of a trace element tends to enter the mineral lattice site normally occupied by a major cation whose chemical properties are similar; for example, where Sr^{2+} enters a Ca^{2+} site. Where differences in physical properties, such as ion size, militate against this, the trace cation tends toward a physically possible site occupied by a major cation whose chemical properties are closest; for example, where Sr^{2+} enters a K^{+} site. Some trace elements are only accommodated with difficulty in the lattice sites of the major mineral-forming elements, and these tend to behave independently; for example, Zr and Cu. Thus, where trace elements have no similar major elements with which they can associate, they tend to concentrate in the magma until they can form a separate mineral phase. Their behavior then switches from that of a trace to that of a major mineral-forming element, with which other suitably matched trace elements may then associate.

Established coherence relationships show that a trace element enters a mineral lattice on a proportional basis relative to its concentration and to that of the major or minor element with which it associates. Coherence relationships based on whole-rock distributions can be misleading unless substantiated in at least one mineral series of the fractionated rock series. The search for element coherence relationships in the fractionated rock and mineral series has revealed the following associations.

(1) Strontium mainly enters feldspar, and in the rock series follows (K + Ca) in the early fractionation stages and K during the later stages. It appears to follow mainly K in plagioclase and Ca in apatite and clinopyroxene.

(2) Barium mainly enters feldspar and biotite. In these, and the rock series, it follows K distribution throughout, whereas in the Fe-rich olivine, the Fe-rich orthopyroxene, and apatite it apparently follows Ca.

(3) The similar geochemical behavior of the transition elements Cr, V, and Sc is consistent with their similar chemical properties. They follow Fe, Ti, and Mn in their coherence relationships in the rock and mineral series, and Cr may also follow Al. They mainly entered clinopyroxene and the opaque iron minerals, and in lesser amounts, orthopyroxene. Site preference in these minerals appears to be largely determined by the achievement of favorable bonding characteristics in the crystal structures. The preferred order of entry into the minerals with fractionation is Cr $>$ Mn$\gtreqless$ V $>$ Ti $>$ Sc $>$ Fe.

(4) The closeness in the chemical properties of Co, Ni, Fe, and Mg is consistent with the similar geochemical behavior shown by Co and Ni. Both entered the Mg-rich olivine freely and show coherence with Mg, whereas in the pyroxenes Co mainly follows Fe^{2+}, and Ni magnesium. Both apparently follow Fe^{2+} in the opaque iron minerals. Cobalt and Ni discriminate between Fe^{2+} and Mg sites according to lattice type, and the preferred order of entry appears to be largely determined by the order of increasing magnitude in the bond energies of the cations in the different structures as other properties influencing their entry appear to be nearly equal. In olivine the preferred order of entry is Ni $>$ Co $>$ Mg $>$ Fe^{2+}, whereas in clinopyroxene and orthopyroxene, the preferred order is Ni $>$ Mg $>$ Co $>$ Fe^{2+}. Thus, it appears that the bond energy of Mg in olivine is weaker than that of Co, whereas the reverse is the case in the pyroxenes. The discrimination shown by Co and Ni for particular lattice sites in olivine and their distribution in this mineral with fractionation demonstrate the significance of site preference in trace element partition equilibrium, and of proportional entry in coherence relationships with a major or minor element. In the rock series Co coherence is with Mg in the early fractionation stages and with Fe^{2+} in the late stages; in the middle stages it is with (Fe^{2+} + Mg). However, for Ni the coherence is with (Mg + Fe^{2+}) in the early stages, and in the middle and late stages it is with Mg.

(5) Yttrium mainly associates with apatite, and either enters Ca sites or forms a separate phosphate phase intergrown with apatite. Apart from this association Y enters the ferromagnesian and opaque iron minerals in

small amounts, and coherence is not with Ca, but with Ti and Fe. In the rock series there appears to be a relationship between Y and K distribution.

(6) The main feature of Zr behavior is that it formed its own silicate phase, zircon, rather than enter other mineral lattices. It does, however, enter some in small amounts, and in the opaque iron minerals and fayalite it appears to follow Ti.

(7) Like Y, lanthanum is largely associated with apatite, where it may also be in some way occluded in mixed phosphate formation. It enters the ferromagnesian minerals in small amounts only, where it follows Ca in early fractionation stages and (Ti + Fe) in the later stages. In the opaque iron minerals, its coherence is with Ti and possibly Fe. In the rock series coherence is with (Ca + K) in the early and middle fractionation stages, and with (Ti + Fe) in the middle and late stages. In addition, a rough relationship with K is apparent throughout.

(8) Copper in the intrusion crystallized mainly as sulfide phases and entered other minerals in small amounts only. Some of that in the minerals probably includes Cu as minute grains of sulfide and adsorbed Cu. It probably follows Na^{+} in plagioclase and Fe^{2+} in the ferromagnesian and opaque iron minerals and in apatite.

Determinations for Nb, Mo, Nd, B, Ga, Sn, and Pb were made mainly to establish their variation with fractionation in the rock series; the investigation was insufficient for a complete study of their geochemical behavior.

REFERENCES CITED

Ahrens, L. H., 1963, The significance of the chemical bond for controlling the geochemical distribution of the elements. Part 1: Phys. and Chem. Earth, v. 5, p. 1-54.

Ahrens, L. H., and Taylor, S. R., 1961, Spectrochemical Analysis: 2nd ed. Addison-Wesley Publ., 454 p.

Beevers, C. A., and McIntyre, D. B., 1945, The atomic structure of fluorapatite and its relation to that of tooth and bone material: Mineralog. Mag., v. 27, p. 254-257.

Bowen, N. L., 1915, Crystallization-differentiation in silicate liquids: Am. Jour. Sci., 4th ser., v. 39, p. 175-191.

Bowen, N. L., and Schairer, J. F., 1935, The system MgO-FeO-SiO_2: Am. Jour. Sci. v. 29, p. 151-217.

Brown, G. M., 1957, Pyroxenes from the early and middle stages of fractionation of the Skaergaard intrusion, Eastern Greenland: Mineralog. Mag., v. 31, p. 511-543.

Brown, G. M., and Vincent, E. A., 1963, Pyroxenes from the late stages of fractionation of the Skaergaard intrusion, East Greenland: Jour. Petrology, v. 4, p. 175-196.

Burns, R. G., and Fyfe, W. S., 1964, Site preference energy and selective uptake of transition-metal ions from a magma: Science, v. 144, no. 3621, p. 1001-1003.

—— 1966, Behavior of nickel during magmatic crystallization: Nature, v. 210, no. 5041, p. 1147-1148.

Carr, M. H., and Turekian, K. K., 1961, The geochemistry of cobalt: Geochim. et Cosmochim. Acta, v. 23, p. 9-60.

Chayes, F., 1952, Notes on the staining of potash feldspar with sodium cobaltinitrite in thin section: Am. Mineralogist, v. 37, p. 337-340.

Darton, N. H., 1890, The relations of the traps of the Newark system in the New Jersey region: U.S. Geol. Survey Bull. 67, p. 37-53.

Deer, W. A., Howie, R. A., and Zussman, J., 1962, Rock-forming minerals: v. 3, Sheet silicates, v. 1, Ortho- and ring silicates: John Wiley & Sons.

—— 1963, Rock-forming minerals: v. 2, Chain silicates: John Wiley & Sons.

Edwards, A. B., 1942, Differentiation of the dolerites of Tasmania: Jour. Geology, v. 50, p. 451-480, 579-610.

Erickson, G. P., and Kulp, J. L., 1961, Potassium-argon measurements on the Palisades Sill, New Jersey: Geol. Soc. America Bull., v. 72, p. 649-652.

Friedman, G. M., 1954, Note on the relative abundance of some trace elements near the lower and upper contacts of the Palisades Sill—A communication: Am. Jour. Sci., v. 252, p. 502-503.

Goldschmidt, V. M., 1937, The principles of distribution of chemical elements in minerals and rocks: Jour. Chem. Soc., p. 655-673.

—— 1944, Crystal chemistry and geochemistry: Chemical Products, v. 7, p. 29-34.

—— 1954, Geochemistry, *in* Muir, A., *Editor,* Oxford University Press, 730 p.

Greenland, L., and Lovering, J. F., 1966, Fractionation of fluorine, chlorine, and other trace elements during differentation of a tholeiitic magma: Geochim. et Cosmochim. Acta, v. 30, p. 963-982.

Gunn, B. M., 1962, Differentiation in Ferrar dolerites, Antarctica: New Zealand Jour. Geology and Geophysics, v. 5, p. 820-863.

—— 1963, Layered intrusions in the Ferrar dolerites, Antarctica: Mineral. Soc. America Spec. Paper 1, p. 124-133.

—— 1966, Modal and element variations in Antarctic tholeiites: Geochim. et Cosmochim. Acta, v. 30, p. 881-920.

Hamilton, W., 1964, Diabase sheets differentiated by liquid fractionation, Taylor Glacier region, *in* Adie, R. J., *Editor,* South Victoria Land (Antarctica); SCAR Cap Town Symposium of Antarctic Geology.

—— 1965, Diabase sheets of the Taylor Glacier region, Victoria Land, Antarctica: U.S. Geol. Survey Prof. Paper, 456-B, 71p.

Hess, H. H., 1941, Pyroxenes of common mafic magmas: Am. Mineralogist, v. 26, p. 515-535, 573-594.

—— 1949, Chemical composition and optical properties of the common clinopyroxenes. Part 1: Am. Mineralogist, v. 34, p. 621-666.

—— 1956, The magmatic properties and differentiation of dolerite sills—A critical discussion: Am. Jour. Sci., v. 254, p. 446-451.

—— 1960, Stillwater igneous complex, Montana: Geol. Soc. America Mem. p. 80, 230.

Jaeger, J. C., and Joplin, G. A., 1955, Rock magnetism and the differentiation of dolerite sill: Australian Geol. Soc. Jour., v. 2, p. 1-19.

Jaeger, J. C., 1957, The temperature in the neighborhood of a cooling intrusive sheet: Am. Jour. Sci., v. 255, p. 306-318.

Joplin, G. A., 1957, The problem of the quartz dolerites: Some significant facts concerning mineral volume, grain size and fabric: Royal Soc. Tasmania Papers and Proc., v. 91, p. 129-142.

Kay, M., 1951, North American geosynclines: Geol. Soc. America Mem. 48, 143 p.

Kummel, H. B., 1897, The Newark system: New Jersey Geol. Survey Ann. Rept.

—— 1900, The Newark or red sandstone rocks of Rockland County, New York: New York State Museum, 2nd Ann. Rept. Regents, 1898, v. 2, p. 9-50.

Kuno, H., 1955, Ion substitution in the diopside-ferropigeonite series of clinopyroxenes: Am. Mineralogist, v. 40, p. 70-93.

Lewis, J. V., 1907, The origin and relations of the Newark rocks: New Jersey Geol. Survey Ann. Rept., 1906, p. 99-129.

—— 1908a, Petrography of the Newark igneous rocks of New Jersey: New Jersey Geol. Survey Ann. Rept., 1907, p. 99-167.

—— 1908b, The Palisade diabase of New Jersey: Am. Jour. Sci., 4th ser., v. 26, p. 155-162.

—— 1915, Origin of the sedimentary minerals of the Triassic trap rocks: New Jersey Geol. Survey Ann. Rept., 1914, Bull. 16, p. 48-49.

Lieberman, K. W., 1966, The determination of bromine in terrestrial and extraterrestrial materials by neutron activation analysis: U.S. Atomic Energy Comm. Tech. Rept. ORO-2670-13, 166 p.

Lowe, K. E., 1959, Structure of the Palisades intrusion at Haverstraw and West Nyack, New York: New York Acad. Sci. Annals, v. 80, p. 1127-1139.

Mason, B. H., 1960, Trap rock minerals of New Jersey: New Jersey Geol. Survey Bull. 64, 51 p.

McDougall, Ian, 1961, Optical and chemical studies of pyroxenes in a differentiated Tasmanian dolerite: Am. Mineralogist v. 46, p. 661-687.

—— 1962, Differentiation of the Tasmanian dolerites: Red Hill dolerite-granophyre association. Geol. Soc. America Bull., v. 73, p. 279-316.

—— 1964, Differentiation of the Great Lake dolerite sheet, Tasmania: Australian Geol. Soc. Jour., v. 11, pt. 1, p. 107-132.

McDougall, Ian, and Lovering, J. F., 1963, Fractionation of chromium, nickel, cobalt, and copper in a differentiated dolerite-granophyre sequence at Red Hill, Tasmania: Australian Geol. Soc. Jour., v. 10, pt. 2, p. 325-338.

Muir, I. D., 1951, The clinopyroxenes of the Skaergaard intrusion, eastern Greenland: Mineralog. Mag., v. 29, p. 690-714.

—— 1954, Crystallization of pyroxenes in an iron-rich diabase from Minnesota: Mineralog. Mag., v. 30, p. 376-388.

Murata, K. J., and Richter, D. H., 1966, Chemistry of the lavas of the 1959-60 eruption of Kilauea Volcano, Hawaii: U.S. Geol. Survey Prof. Paper, 537-A, 26 p.

Nockold, S. R., 1954, Average chemical compositions of some igneous rocks: Geol. Soc. America Bull., v. 65, p. 1007-1032.

Nockolds, S. R., and Allen, R., 1956, The geochemistry of some igneous rock series: III: Geochim. et Cosmochim. Acta, v. 9, p. 34-77.

Osborne, E. F., 1959, Role of oxygen pressure in the crystallization and differentiation of basaltic magma: Am. Jour. Sci., v. 257, p. 609-647.

Pauling, L., 1960, The nature of the chemical bond, 3rd ed.: New York, Cornell Univ. Press, 644 p.

Poldervaart, Arie, 1944, The petrology of the Elephant's Head dike and the New Amalfi sheet (Matatiele): Royal Soc. South Africa Trans., v. 30, p. 85-119.

Poldervaart, Arie, and Hess, H. H., 1951, Pyroxenes in the crystallization of basaltic magma: Jour. Geology, v. 59, p. 472-489.

Poldervaart, Arie, and Walker, K. R., 1962. The Palisades Sill: Internat. Mineral. Assoc., 3rd. Gen. Cong. Northern Field Excursion Guidebook.

Powers, H. A., 1955, Composition and origin of basaltic magma of the Hawaiian Islands: Geochim. et Cosmochim. Acta, v. 7, p. 77-107.

Ramdohr, P., 1953, Ulvospinel and its significance in titaniferous iron ores. Econ. Geology, v. 48, p. 677-688.

Ribbe, P. H., and Smith, J. V., 1966, X-ray emission microanalysis of rock-forming minerals, IV. Plagioclase feldspars: Jour. Geology, v. 74, p. 197-233.

Richter, D. H., and Moore, J. G., 1966, Petrology of the Kilauea Iki lava lake, Hawaii: U.S. Geol. Survey Prof. Paper 537-B, 26 p.

Richter, D. H., and Murata, K. J., 1966, Petrology of the lavas of the 1959-60 eruptions of Kilauea Volcano, Hawaii: U.S. Geol. Survey Prof. Paper 537-D, 12 p.

Ringwood, A. E., 1955a, The principles governing trace element distribution during magnetic crystallization. Part I: The influence of electronegativity: Geochim. et Cosmochim. Acta, v. 7, p. 189-202.

—— 1955b, The principles governing trace element behavior during magmatic crystallization. Part II: The role of complex formation: Geochim. et Cosmochim. Acta, v. 7, p. 242-254.

Schairer, J. F., Smith, J. R., and Chayes, F., 1956, Refractive indices of plagioclase glasses: Carnegie Inst. Washington Year Book, no. 55, p. 195-197.

Shaw, D. M., 1960, Spectrochemical analysis of silicates using the Stallwood jet: Canadian Mineralogist, v. 6, p. 467-482.

Simkin, T., 1967, Flow differentiation in the pictritic sills of North Syke: *in* Wyllie, P. J., *Editor*: Ultramafic and related rocks, John Wiley & Sons.

Taylor, S. R., 1966, The application of trace element data to problems in petrology: Phys. and Chem. Earth, v. 6, p. 133-213.

Taylor, S. R., and Kolbe, P., 1964, Geochemical standards: Geochim. et Cosmochim. Acta, v. 28, p. 447-454.

Thompson, H. D., 1959, The Palisades ridge in Rockland county, New York: New York Acad. Sci. Annals., v. 80, p. 1106-1126.

Turekian, K. K., Gast, P. W., and Kulp, J. L., 1957, Emission-spectrographic method for the determination of strontium in silicate materials: Spectrochimica Acta, v 9, p. 40-46.

Vogt, J. H. L., 1923, On the content of nickel in igneous rocks: Econ. Geology, v. 18, p. 307-352.

Wager, L. R., and Deer, W. A., 1939, The petrology of the Skaergaard intrusion, Kangerdlugssuaq, East Greenland: Medd. om Grønland, bd. 105, no. 4, 335 p.

Wager, L. R., and Mitchell, R. L., 1951, The distribution of trace elements during strong fractionation of basic magma—A further study of the

Skaergaard intrusion, East Greenland: Geochim. et Cosmochim. Acta, v. 1, p. 129-208.

Wager, L. R., Brown, G. M., and Wadsworth, W. J., 1960, Types of igneous cumulates: Jour. Petrology, v. 1, p. 73-85.

Walker, F., 1940, The differentiation of the Palisade diabase, New Jersey: Geol. Soc. America Bull., v. 51, p. 1059-1106.

—— 1952, Late magmatic ores and the Palisade diabase sheet: Econ. Geology, v. 47, p. 349-351.

—— 1953, The pegmatic differentiates of basic sheets: Am. Jour. Sci., v. 251, p. 41-60.

—— 1956, The magmatic properties and differentiation of dolerite sills—A critical discussion: Am. Jour. Sci., v. 254, p. 433-443.

—— 1957, Ophitic texture and basaltic crystallization: Jour. Geology, v. 65, p. 1-14.

Walker, F., and Poldervaart, Arie, 1949, Karroo dolerites of the Union of South Africa: Geol. Soc. America Bull., v. 60, p. 591-706.

Walker, K. R., 1965, Re-examination of the Palisades Sill: International Council Scientific Unions, Upper Mantle Project, Australian Prog. Rept.

SUBJECT INDEX

AUTHOR INDEX